GRAND UNIFIED THEORY MADE EASY

Third edition

Charles R. Storey

Copyright © 2015

published by

Copyright © 2015 by Charles R. Storey

all rights reserved

including the right of reproduction

in whole or in part in any form

ISBN 0-9638766-5-1

Printed in the United States of America

GRAND UNIFIED THEORY MADE EASY

Third edition

Charles R. Storey

Copyright © 2015

DEDICATION

This book is dedicated to all the physicists, assistants, undergraduates, technicians, and secretaries around the world whose hard work and dedication has contributed to the advancement of physical science.

Forward

Many publications on this subject matter rely upon mathematics, descriptions and terminology which can be very technical and difficult to follow. In this book, you will find many of the descriptions greatly simplified and easier to understand. After you have read several chapters, you may come to the conclusion that the same underlying physical process is being described over and over again. This book is written in this manner to demonstrate that many of our laws of physics, including the principles of electricity, magnetism, and gravity, are a result of the same, relatively simple, underlying physical process which is producing our perceptions about our universe in general and the world in which we live.

CONTENTS

		page
Forward		v
Acknowledgements		vii
Introduction		viii
Chapter 1	OUR UNIVERSE BEGINS	11
Chapter 2	UNDERSTANDING RELATIVITY	25
Chapter 3	GRAVITATIONAL, ELECTRIC, AND MAGNETIC FIELDS	49
Chapter 4	ELECTROMAGNETIC WAVES IN FOUR-DIMENSIONAL SPACE-TIME	84
Chapter 5	UNDERSTANDING PLANCK'S CONSTANT	114
Chapter 6	THE FUTURE OF SPACE EXPLORATION	147
Chapter 7	THE FUTURE OF OUR UNIVERSE	168
Glossary		187
Bibliography		195
Appendix		198

ACKNOWLEDGEMENTS

The scientists, mathematicians, theoretical physicists, and assistants whose determination and hard work have made this unified theory possible, deserve our sincerest appreciation and thanks

As the information in this publication was being prepared, it was possible to gain some insight regarding the long hours of hard work, the disappointments, the triumphs, and the personal sacrifices involved in scientific research.

INTRODUCTION

The original goal in the development of this theory was to discover some of the connections between the primary forces of nature, including electricity, magnetism, gravity, the weak nuclear force, and the strong nuclear force. A considerable amount of reliable experimental evidence has already been compiled throughout the years, which provides us with a wealth of information that we may, someday, be able to piece together as a computer model.

By the latter half of the nineteenth century, starting with James Clerk Maxwell, a number of great minds were working diligently to discover the nature of a mysterious physical substance, which, they suspected, was conducting light waves through space, and which was causing the effects of electricity and magnetism.

An investigation into some of the works which were produced during this time period has revealed that the Permittivity and Permeability "constants" (these will be described in greater detail later) might be a good place to start looking for more information regarding the characteristics of this mysterious physical substance, if it exists at all. Also, clues regarding the nature of this substance were turning up in the mathematics which is why so many of these great physicists were strongly convinced of its existence in one form or another.

Once it was suspected that this mysterious physical substance might possibly be involved in an underlying physical process which we are not able to observe directly, much of the information which had been collected previously began falling into place.

The considerable amount of information which was assimilated concerning Relativity, electromagnetic waves, and Quantum Mechanics, then began to merge into a much more simplified overview which does not seem to be that complicated or difficult to understand.

The mathematics which are included, are there mainly just to show you the patterns and symmetries in these different relationships. Initially, you may be surprised at how some of these physical processes fit together. This theory also represents a challenge to other physicists, both professional and amateur alike, to use this information to make even more discoveries regarding the underlying physical process which produces the functioning of our observed universe in addition to atomic interactions.

This information is being presented along with some experimental evidence to back it up as well as some experiments which you can perform in order to determine for yourself whether or not this approach is correct. So you be the judge, and may you receive many wonderful hours of enjoyment in reaching your conclusions regarding this surprising and interesting field of science.

CHAPTER 1

OUR UNIVERSE BEGINS

One of the greatest challenges which lies ahead of us in physical science, is to develop a "Grand Unified Theory", or, "Theory of Everything", which will explain our observations in our universe on the astronomical scale, and everything we have observed on the atomic scale. Taking together the experimental and observational evidence which has accumulated up until the present time and then fitting this information together into a credible unified theory, represents a formidable challenge which our national and international scientific community has been working on for quite some time.

The definition of a theory is: "A hypothesis, or exposition of the abstract principles of a science or art, as opposed to practice". Any hypothesis which has been developed to describe observed cosmological events, or other universal physical processes, is just a theory, and no theory can be proved to be 100% correct. As with any theory, more detailed information will eventually be included to account for additional experimental and observational evidence which will be discovered later on.

Let us start by trying to develop a clearer picture of how our universe, and our solar system, came into existence. In order to do this, we shall need to decide which theories, observations, and experimental evidence we can use to accomplish this objective. Then, we can arrange this information into a format which will describe the observations and experimental results in a logical, step-by-step manner that can make it easy for us to read and understand.

Due to the extremely large quantity of information which has been previously discovered, recorded, and published in each of these areas, and the overall scope of this scientific overview, much of the supporting experimental evidence, and several of these theories, are mentioned here only briefly. Much more detailed information on each of these related topics is readily available to the reader in works devoted to these various subjects and in published experimental results which are available in libraries, or over the Internet.

The first major theory which we will need to use in order to develop this sequence of observational and experimental evidence is "A Treatise on Electricity and Magnetism", actually a collection of theories, which was first published in 1873 by James Clerk Maxwell (1831 – 1879). This collection of theories was instrumental in laying the groundwork for much of our present-day understanding of electricity and magnetism.

Most importantly, it predicted the existence of electromagnetic waves, and it also predicted, with considerable accuracy, the speed at which these waves would propagate (travel) through space. This challenged physicists to conduct further experimentation with electromagnetic waves (light), which, in turn, led to more new discoveries, and ultimately, to the development of our next major theory.

This second important theory is "The Special Theory of Relativity" which Albert Einstein (1879 – 1955) introduced in 1905. Some of the experimental evidence which led to the development of the Special Theory of Relativity was provided by the previous research efforts of James Clerk Maxwell, Albert A. Michelson (1852 – 1931), Hendrik A. Lorentz (1853 – 1928), and several other prominent physicists of that particular time period. The Special Theory of Relativity provides us with some of the most important physical relationships and mathematics which are used to describe the basic underlying physical processes which produce our perceptions of reality and our universe in general.

The third theory is the "Theory of an Expanding Universe". This theory has a considerable number of observations to back it up. Evidence supporting this theory was first discovered by American astronomer Edwin Powell Hubble (1889 – 1953). In 1919, he joined the staff of the Mount Wilson Observatory at the

University of California, where he did most of his work as an astronomer.

Using the 100-inch telescope on top of Mount Wilson, California, Dr. Hubble began to study objects which were "diffuse" (fuzzy), and, therefore, could not be identified as individual stars. It was soon discovered that some of these objects were glowing clouds of gas and dust, now known as nebulae.

Some of these strange objects turned out to be comprised of very large numbers of individual stars. Later, these large clusters of stars came to be known as galaxies. In 1924, Dr. Hubble's distance calculations proved conclusively that these distant galaxies were not part of the structure of our own Milky Way galaxy.

The motion of a star or galaxy, either toward or away from us, can be determined by analyzing the spectrum of light it emits. If a star is moving toward us, the frequency of its light spectrum is shifted toward the blue, or, higher frequency end of the spectrum. If it is moving away from us, then its light spectrum is shifted toward the red, or lower frequency, end of the spectrum. This is known as the "Doppler Effect".

As more and more galaxies were discovered, the spectrum of light they emit indicated that there was a red-shift in all of the galaxies. In other words, large numbers of galaxies are moving away from us in all directions (receding), with the more distant galaxies moving away at extremely high velocities.

(Continued on page 16)

LORENTZ TRANSFORMATION EQUATIONS

Graph of the length equation: $\quad L' = L\sqrt{1 - \frac{v^2}{c^2}}$

L = Length;   L' = Distorted Length;   v = velocity;   c = speed of light

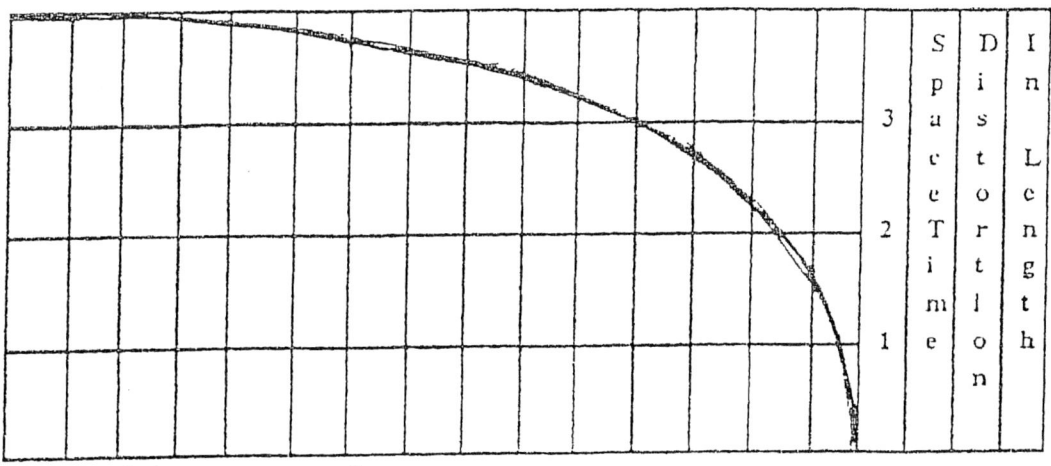

← "where we are now"

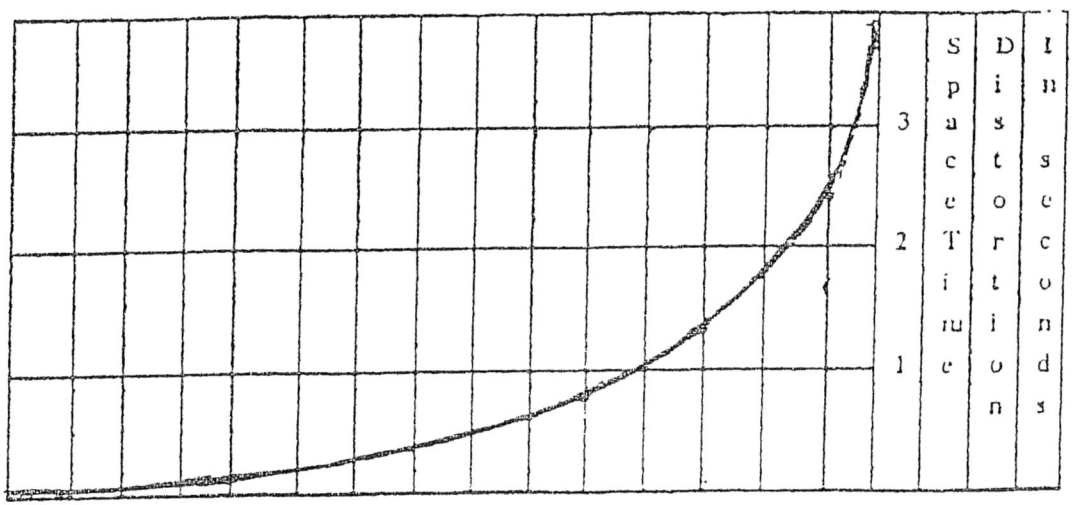

Speed of light " c

"←"where we are now"

Graph of the time equation: $\quad T' = \dfrac{T}{\sqrt{1 - \frac{v^2}{c^2}}}$

T = Time;   T' = Distorted Time;   v = velocity;   c = Speed of light

(Continued from page 14)

This discovery led to the conclusion that approximately 15 billion years ago, our universe, as we know it, originated from a single point in otherwise empty space and began expanding very rapidly. This observational evidence eventually led to the development of the cosmological "Big Bang Theory" which was introduced in 1964.

Surprisingly, in the early 1930's, it was also discovered that the graphs of the Lorentz time and length equations, which Albert Einstein proposed as a mathematical description of our space-time continuum in his Special Theory of Relativity, also appear to provide us with a clear and simple mathematical description of this physical expansion process when solved for greater and greater velocities.

The graphs on page 15 are an excellent example of how physicists have been able to use mathematics as a kind of "scientific short-hand" to connect the various numerical values, mathematical relationships, and graphs which are able to provide us with a clear and precise description of how all the physical processes in our entire universal system fit together. As we shall see later on, a large number of important new scientific discoveries have been made by analyzing numerical patterns, graphs, and the mathematical relationships of different physical processes.

The two graphs on page 15, when placed together, appear to describe the pressure-release of some type of "physical substance" into empty space. As the size of the cloud of this substance is expanding and becoming larger (upper graph), the pressure or concentration of it is going down in direct proportion (lower graph).

The theory that our universe originated in a gigantic explosion (a big bang) was first introduced by a Russian-American theorist, George Gamow (1904 – 1968), in 1964. As our technology continued to advance and additional experimental and observational evidence continued to accumulate, the Big Bang Theory began to emerge as the only cosmology theory capable of explaining all, or almost all, of the results of recent astronomical observations.

The Big Bang theory is also supported by the discovery of the cosmic microwave background radiation, discovered in 1975, by Robert W. Wilson (1936 -   ) and Arno A. Penzias (1933 -   ), working at Bell Labs in New Jersey. The degree of redshift of this radiation, from the visible light frequencies into the microwave frequency region (see pages 93 and 94), would indicate that the outer reaches of our universe are expanding in all directions, like an expanding sphere, at extremely high velocities.

Further analysis of this cosmic microwave background radiation has demonstrated, conclusively, that it is essentially a visible light spectrum (see pages 117 and 118) which has been red-shifted all the way down into the microwave frequency range by the extremely high recessional velocity of the outer reaches of our universe (more on this in chapter 5).

A number of additional details regarding this Cosmic Microwave Background (CMB) radiation have confirmed other important astronomical events regarding the formation of our early universe in addition to some of the earlier predictions made by the "Big Bang Theory". Once again, we can never be absolutely certain that any theory is 100% correct, but at the moment, the "Big Bang Theory" seems to agree with our observations better than any other cosmology theory that we have at the present time.

An overview of the "Big Bang Theory" would reveal that our universe started from a single point in otherwise empty space and has expanded outward in all directions, at a very high velocity for approximately 15 billion years. This accounts for the vastness of our universe today.

Taken together with James Clerk Maxwell's "Treatise on Electricity and Magnetism" and Albert Einstein's "Special Theory of Relativity", we have experimental and observational evidence which suggests that our universal expansion process is apparently being driven by the "pressure-release" of a mysterious "physical substance" which we are not able to observe directly (see page 15).

We are only able to observe the "effects" of this "pressure-related" expansion process and it is these "effects" which are producing our observations (perceptions) of an expanding universe and also our perceptions regarding our four-dimensional space-time continuum. Albert Einstein introduced the concept of "four-dimensional space-time" when he presented his Special Theory of Relativity in 1905.

In Einstein's view, we live in a four-dimensional universe where the three coordinates of space; height, length, and width (size), are connected with, and inversely proportional to the passage of time. This means that, as experiments have shown, when you have an increase in relative motion between two objects, time slows down so that one second lasts longer, or becomes greater, and the three dimensions of space contract, or become smaller with respect to each object.

Experiments have also demonstrated that space and time are so closely related, in fact, that you cannot change one without affecting the other. This close relationship is revealed by the Big Bang Theory in stating that our universe began at a single point in space, approximately fifteen billion years ago, and has been expanding outward in all directions ever since, like an expanding sphere (see page 20).

This is a fundamental concept which not only describes things on an astronomical scale, but, also, many things on the atomic (quantum) scale as well. A sphere is defined as a set, or group, of all points in space which are a certain distance from the center point (point of origin). Many of the events involving four-dimensional space-time can be described by the surface of a sphere expanding outward in all directions (three dimensional) at once. The three dimensions of space describe the "distance" in height, length, and width, or the size of the sphere. The number of seconds describes the fourth dimension which is time, or rate of expansion.

(Continued on page 21)

SPACE CHART

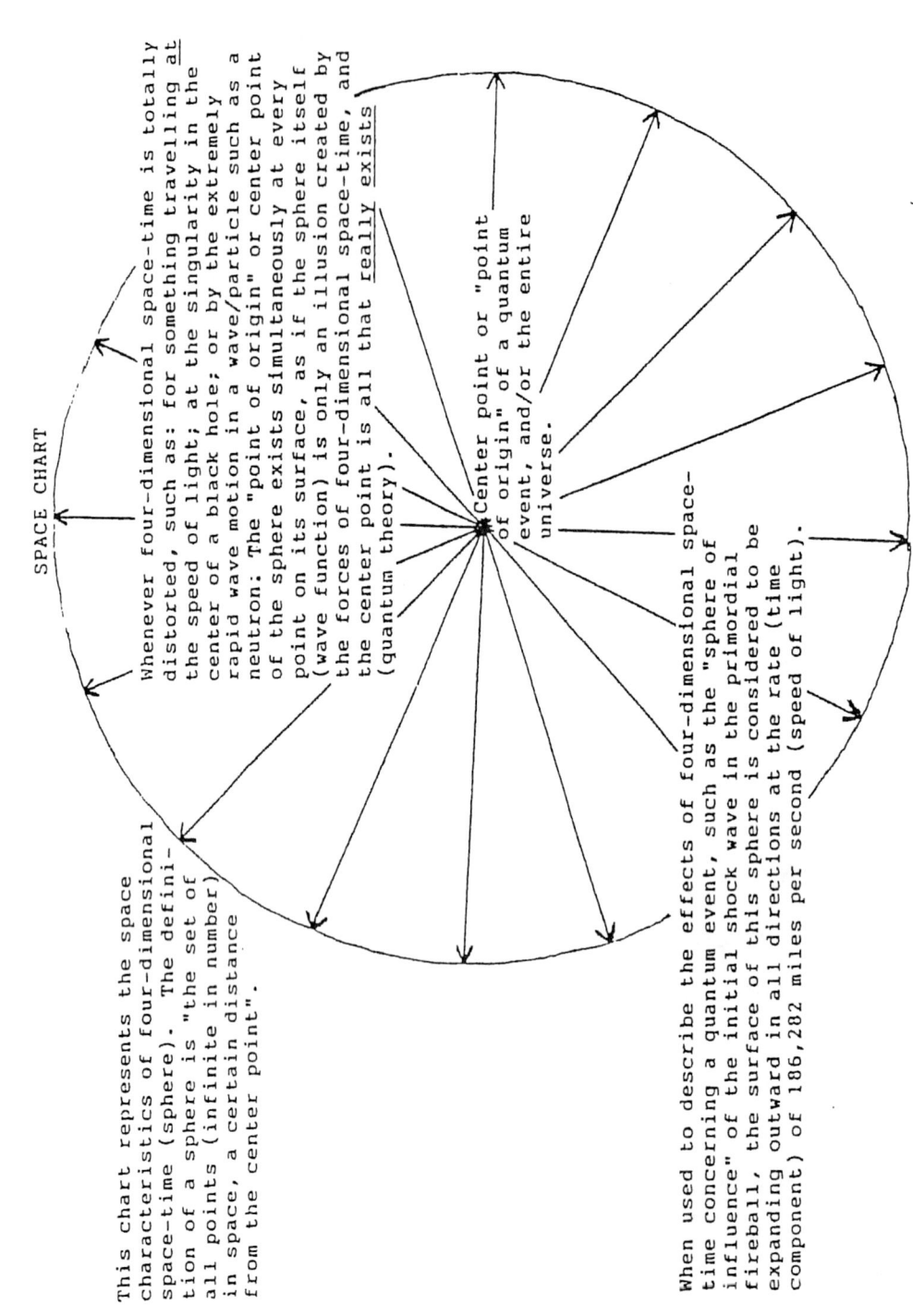

SPACE CHART

This chart represents the space characteristics of four-dimensional space-time (sphere). The definition of a sphere is "the set of all points (infinite in number) in space, a certain distance from the center point".

Whenever four-dimensional space-time is totally distorted, such as: for something travelling at the speed of light; at the singularity in the center of a black hole; or by the extremely rapid wave motion in a wave/particle such as a neutron: The "point of origin" or center point of the sphere exists simultaneously at every point on its surface, as if the sphere itself (wave function) is only an illusion created by the forces of four-dimensional space-time, and the center point is all that really exists (quantum theory).

Center point or "point of origin" of a quantum event, and/or the entire universe.

When used to describe the effects of four-dimensional space-time concerning a quantum event, such as the "sphere of influence" of the initial shock wave in the primordial fireball, the surface of this sphere is considered to be expanding outward in all directions at the rate (time component) of 186,282 miles per second (speed of light).

(Continued from page 19)

This is demonstrated by the fact that, as our universe continues to expand, the actual physical expansion process appears to be occurring evenly and consistently throughout our whole universe. It is like our entire universal space-time continuum is continually being "stretched" evenly and proportionately across the whole system as this expansion process continues.

This situation also provides us with important scientific evidence which strongly suggests that there is some type of a "physical substance" which is actually the driving force behind our universal expansion process instead of just the weight of the total mass distribution within our universe and subsequent gravitational attractions. If the inertia of the total mass distribution within our universe, influenced by gravitational attractions, was the force that is really driving the expansion process, then the shape of the curves on the graphs on page 15 would have a sharp spike at or near the starting point and then taper off at a rate described by the inverse of the curves on the two graphs.

The curves on the graphs on page 15 demonstrate conclusively that the physical process (force), which is actually driving our universal expansion process, is just a simple "pressure release" of a mysterious physical substance which we are unable to observe directly, and not the observed mass-related thermal expansion (explosion) process which has occurred as an end-result. According to this theory, as the point (primordial singularity) from which our universe originated, began to expand rapidly, the full-cycle electromagnetic shockwave which forms the outer surface of the primordial singularity, which also represents the outer surface of our universe, began to produce large numbers of separate, relative motion-related, electromagnetic wave identities (neutrons) which then became the progenitors of all the mass in our universe (more on this in chapter 5).

The neutrons, being the first elementary full-cycle electromagnetic wave/particles, eventually began to decay (break apart) into electrons and protons. The electrons and protons then recombined, as seen in laboratory experiments, to form hydrogen atoms. Some of the free neutrons were then "captured" in the nuclei of the hydrogen atoms in order to form helium.

As our early universe continued to expand, large clouds of this newly-formed hydrogen and helium gas began to condense under gravitational pressure and started forming stars and galaxies. Many of the new stars within the earliest galaxies, began their existence as binary systems which consist of two stars rotating around a common center of gravity (approximately two thirds of the stars in our Milky Way Galaxy exist in this manner).

Billions of years later, as many of the larger stars in these binary systems burned out and then exploded as novae and supernovae, the resulting fragmented concentrations of heavier elements then became the planets and moons orbiting the smaller remaining binary partner stars to form our first solar systems.

According to this theory, approximately five billion years ago, our Sun was part of such a binary system in which its larger partner burned up its hydrogen and helium fuel much more rapidly than the Sun (which was much smaller) and then exploded as a supernova. Much of the lighter gaseous materiel of this larger star was blasted away into the outer reaches of our solar system.

The remaining concentrations of heavier elements, which were also fragmented by the explosion, began to form the various planets and moons of our solar system including our planet Earth. Due to the fact that almost all of the planets and moons within our solar system orbit in the same direction and in the same plane (rotational energy is conserved); that there is volcanic activity and great amounts of heat inside the Earth as well as within some of the other planets and moons; and the fact that uranium is found naturally on the Earth in significant quantities; tend to support this hypothesis. Uranium is the heaviest element produced by a star, which is at least two and a half times as massive as our Sun, before it explodes as a supernova.

In 1987, a star which was located in the Large Magellanic Cloud, our closest neighboring galaxy, underwent a fairly large supernova explosion. By observing these types of cosmic events, and by comparing these observations with what we see going on within our own solar system, we will continue to learn more of the specific details regarding the formation of our early solar system.

As more and more of the pieces to this great puzzle are being discovered, it becomes easier and easier to see how they all fit together. In the next few chapters, we shall see how new discoveries, new theories, and new experimental evidence helps physicists and astronomers take bold, new steps toward a better understanding of the universe in which we live.

CHAPTER 2

UNDERSTANDING RELATIVITY

Many years ago, physicists discovered that light behaves, in some respects, like waves which travel across great distances through the vacuum of space to arrive here from our Sun and the stars. Due to the fact that waves need a conducting medium or physical substance of some kind through which to travel, physicists determined that there must be some kind of mysterious, invisible substance in outer space which is conducting the light waves. They named this suspected physical substance "lumeniferous (light carrying) ether".

James Clerk Maxwell, in his "Treatise on Electricity and Magnetism" (1873), provides us with his reasoning and supporting mathematical evidence regarding calculated propagation rates, and the suspected conducting media for his proposed electromagnetic waves.

These mathematics produced, by several different methods, what appeared to be two universal "spring constants": the "Permittivity of Free Space" (the ability of free space to hold an electric charge); and the "Permeability of Free Space" (the ability of free space to hold a magnetic field), which apparently facilitates the action of electric and magnetic fields in addition to the propagation of electromagnetic waves through space (see page 30).

ε_0 = the Permittivity of free space: 0.000,000,000,008,854,187,817,6 Farads per meter

μ_0 = the Permeability of free space: 0.000,000,125,663,706,144 Henry's per meter^{-1} (The $^{-1}$ exponent means that there is an inverse relationship between ε_0 and μ_0)

At the very end of his "Treatise on Electricity and Magnetism", James Clerk Maxwell challenges those who would come after him, to conduct experimentation designed to reveal the existence of this suspected medium which conducts the light waves. He states that: "If we admit this medium as an hypothesis, I think it ought to occupy a prominent place in our investigations, and that we ought to endeavor to construct a mental representation of all the details of its action, and that has been my constant aim in this treatise".

It was this challenge which led to the Michelson-Morley experiment in 1887, and ultimately, to the introduction of the "Special Theory of Relativity" in 1905. Albert Einstein wrestled extensively with the question of this suspected medium, or "physical substance" (see appendix).

In the 1880's, physicists and astronomers knew that the Earth moves at a velocity of approximately 30 kilometers per second in its orbit around the Sun. They concluded that since the Earth is moving through this suspected "ether" medium at such a high velocity, then there should be a sufficient amount of relative motion between the Earth and this suspected "ether" substance to bend the light waves slightly.

In 1887, Albert A. Michelson, a physics professor at the Case School of Applied Science, in Cleveland, Ohio, and Edward W. Morley (1838 – 1923), a chemistry teacher at the Western Reserve University, also in Cleveland, Ohio, began working on a project to prove the existence of this "ether" substance. They built a device which would be able to measure, to within one part in 100 million, the movement of light waves in different directions with respect to the Earth's movement around the Sun.

This is how their experiment worked: They would direct a beam of light toward a thinly silvered mirror which would split the beam, allowing half of the light to pass right on through, and then reflecting the other half toward another mirror. The two separate beams of light would be reflected back and forth in different directions between several mirrors before being brought back together and then directed into an eyepiece.

If the light waves from both beams arrived at the eyepiece slightly out of phase, or blurred, this would indicate that there was a relative motion between the Earth and the "ether substance". To their great surprise, however, the beams of light always arrived at the eyepiece in-phase.

Regardless of the direction in which the experiment was conducted, the light beams always arrived back at the eyepiece in phase. Thus, it became obvious that there is no relative motion between the "ether" in space, and the Earth as it moves in its orbit around the Sun.

Physicists all over the world were astonished by the results of this experiment. Prior to this experiment, they had been almost certain that there had to be some type of relative motion between the Earth and this mysterious "ether" substance which they suspected as being the medium which was conducting the light (electromagnetic) waves through space.

Albert Einstein took a particular interest in the results of this experiment. He used the results of the Michelson-Morley experiment as evidence that the speed of light is constant, but that "space" (size) and the passage of "time" are variable. This led to the development of Albert Einstein's "Special Theory of Relativity". Albert Einstein then developed a much more detailed relativity theory, entitled "The General Theory of Relativity", which he introduced in 1916.

Perhaps the best way to describe the effects of relativity is to use the well-known analogy called the "Twin Paradox". The two twins are "Bill" and "Sam", and Sam is an astronaut. Sam leaves in a spaceship which is headed for a star approximately 10 light years away, while Bill stays back here on Earth.

If Sam's spaceship travels at a velocity which is close to the speed of light, relative to the Earth, then according to Bill, the round trip would require approximately 20 years. As Sam's spaceship accelerates, time onboard the spaceship begins to slow down and the distance to the star "appears" closer to Sam according to the time and length equations in the Special Theory of Relativity (see page 15).

When Sam returns to Earth, his chronometer shows that only two and a half years have passed since he left. However, when Sam sees his brother Bill again, to his surprise, Sam discovers that Bill is now seventeen and a half years older than he is.

How could this be true? And what strange forces could have caused this to happen? First, we can begin to answer these questions by saying that numerous scientific experiments have been conducted to study the effects of relativity, and those experimental results consistently agree that the Theory of Relativity is correct.

If this is the first time that you have been made aware of the effects of relativity, then you must be as shocked and surprised as the entire scientific community all over the world was when Albert Einstein first introduced his "Special Theory of Relativity".

Albert Einstein proposed the hypothesis that the speed of light (electromagnetic waves) remains constant, or "fixed", but that time and space are relative (variable).

(Continued on page 31)

TIME AND LENGTH EQUATIONS IN THE SPECIAL THEORY OF RELATIVITY

For the "time" equation (see page 15) we have:

$$T' = \frac{T}{\sqrt{1 - \frac{v^2}{c^2}}}, \text{ Then substitute } (a \times T)^2 \text{ for } v^2$$

Since $c = \frac{1}{\sqrt{\varepsilon_0 \mu_0}}$, you can substitute $\frac{1^2}{\varepsilon_0 \mu_0}$ for c^2

This gives: $T' = \dfrac{T}{\sqrt{1 - \dfrac{(a\,T)^2}{\varepsilon_0 \mu_0}}}$ for the "time" equation.

Likewise, for the length equation (see page 15) we have:

$$L' = L\sqrt{1 - \frac{v^2}{c^2}}, \text{ Then substitute } (a \times T)^2 \text{ for } v^2$$

Since $c = \frac{1}{\sqrt{\varepsilon_0 \mu_0}}$, you can substitute $\frac{1^2}{\varepsilon_0 \mu_0}$ for c^2

This gives: $L' = L\sqrt{1 - \dfrac{(a\,T)^2}{\varepsilon_0 \mu_0}}$ for the "length" equation.

T' = distorted time
L' = distorted length
T = time in "flat" space (away from any strong gravitational fields).
L = length in "flat" space (away from any strong gravitational fields).
c = speed of light in a vacuum
a = acceleration
v = velocity
T = time in "flat" space (away from any strong gravitational fields).
ε_0 = Permittivity of free space: 8.854,187,8E-12 Farads per meter
μ_0 = Permeability of free space: 1.256,637,061,44E-7 Henry's per meter^{-1} (The $^{-1}$ exponent means that μ_0 is inversely related to ε_0)

(Continued from page 29)

The first question that we need to ask is, "Why would any electrical, or physical values, or anything else within our universe, have a tendency to remain in a "fixed state", when, according to our astronomical observations, we "appear" to live in a universe which is totally dynamic, or constantly changing and expanding?"

This situation clearly indicates that there must be some underlying physical process which is producing the stationary, or "fixed", effects of electricity and magnetism, while, at the same time, it is also producing the effects of relativity (our space-time continuum) which are dynamic (see page 15).

In the modified time and length equations on page 30, you can see that all of the key values -- acceleration, time, length, and the Permittivity and Permeability of free space, are inter-connected in one set of equations.

By mentally comparing these relationships, you can now visualize what will happen if the Permittivity and Permeability values (ε_0 and μ_0) do change, say, in a gravitational field, for instance (see pages 71, 72, 73).

When ε_0 and μ_0 are changed, or altered, in a gravitational field, for instance, that causes T' and L' to change also. Consequently, that is what produces the acceleration that you feel in a gravitational field in addition to "gravitational redshift" and the "lensing" of light waves which has also been observed.

The first confirmation of the Theory of Relativity came during the solar eclipse of 1919. Sir Arthur S. Eddington (1882 – 1944) traveled to an Island off the west coast of Africa, which would be directly in the path of an upcoming solar eclipse. During the eclipse, he conducted observations of several stars, which, at that particular time, could be seen just beyond the surface of the Sun.

The observed locations of the stars were recorded and compared with their known locations later. Sure enough, our Sun's gravitational field had bent or "lensed" the light coming from the stars by the amount predicted in Albert Einstein's theory of relativity. This proves that a gravitational field is the same type of distortion in our four-dimensional space-time continuum as that which is produced by the acceleration of an object to give us our "perceptions" of relative motion.

Another test of the relativity theory came in 1971 by Joseph C. Hafele (1933 – 2014), an American physicist, and Richard E. Keating (dates not available), an American Astronomer. They carried out their experiment by taking four very precise atomic clocks around the world in commercial airliners. One trip was made going eastward, and the other was made going westward, with both journeys taking about three days.

The eastbound clocks lost about 59 billionths of a second while the westbound clocks gained 273 billionths of a second. The results agreed very closely with the predictions which were made by the Theory of Relativity.

In September 1975, the Theory of Relativity was tested once again with two "ensembles" of very precise atomic clocks, one on the ground, and one in a U. S. Navy aircraft. The aircraft was flown slowly in a 30-kilometer loop around and around the Chesapeake Bay for fifteen hours at an altitude of approximately 9,000 meters.

Laser flashes every three minutes compared the time intervals recorded in the air and on the ground. Radar tracked the aircraft continuously to compensate for its relative motion. The results were conclusive, the ground-based clocks were running about 50 billionths of a second slower than the clocks on the plane, proving that time slows down as the strength of a gravitational field increases.

Other, more precise experiments have been conducted since then, all of which are stunningly conclusive that Albert Einstein's Theory of Relativity provides us with an accurate mathematical description of how our universal four-dimensional space-time continuum is producing our perceptions of time, distance, relative motion, gravity, energy, and mass.

Most importantly, the underlying message of the Theory of Relativity is that our space-time continuum is comparable to a huge, four-dimensional hologram which produces all of our perceptions and observations about our universe and our world in which we live.

From the mathematical relationships on page 30, we can see that all of the values in the speed of light equation

$\left(c = \frac{1}{\sqrt{\varepsilon_0 \mu_0}}\right)$ are supposedly "constants", and the Permittivity and Permeability values (ε_0 and μ_0), determine the rate at which electromagnetic (light) waves propagate through our space-time continuum.

The Permittivity of free space (ε_0) is a value representing the ability of free space to contain an electric charge. The Permeability of free space (μ_0) is a value representing the ability of free space to contain a magnetic field. Taken together, the magnitude of these values determines the relative strength (effects) of our space-time continuum at each point throughout universal space.

Since we know that electromagnetic waves (light) are "lensed" in a gravitational field (in the presence of atoms), this reveals to us that electricity, magnetism and, gravity are very closely related and also that the magnitude of the Permittivity and Permeability values are reduced, or weakened, in a gravitational field. That is what reduces the frequency of oscillation of electromagnetic waves in addition to the slight changes in the propagation rate which causes the bending, or lensing, of the electromagnetic (light) waves in a gravitational field according to the mathematical relationships on pages 15 and 30).

ε_0 the Permittivity of free space: 8.854,187,817,6E-12 Farads per meter

μ_0 the Permeability of free space: 1.256,637,061,44E-7 Henry's per meter^{-1} (The $^{-1}$ exponent means that there is an inverse relationship between ε_0 and μ_0)

Now, we are provided with a very important common denominator to work with here because the Permittivity (ε_0) and Permeability (μ_0) values (represented by " c" the speed of light) are used in practically all of the mathematics which are used to describe the "fixed" effects of electricity and magnetism, the "dynamic" effects of relativity (velocity and energy), and also atomic electromagnetic wave relationships.

Since the Permittivity and Permeability constants are not really that "constant" after all, then this strongly suggests that the relative strengths of the Permittivity and Permeability gradients, from an area of higher values to an area of lower values, are what is producing the acceleration that we experience in a gravitational field; the accelerations associated with relative motion; and also the accelerations between electric and magnetic fields (see page 73).

The Permittivity and Permeability values may require further study for us to better-understand them, and quite possibly, we need to develop a measuring device which would better-enable us to take accurate measurements at different levels within the Earth's gravitational field, and with varying degrees of relative motion.

The continuing and ongoing expansion process, described by the relativity equations, is actually the great engine of our universe. It is making time pass as we know it. It keeps the stars burning to produce energy and to maintain the strength of our gravitational fields and the forces which hold the atoms together.

Moreover, the "effects" of this relativistic energy field, which was named "the four-dimensional space-time continuum" by Albert Einstein, causes our universe and everything in it to "appear" to us as though it is spatially extended to its present size, and to "appear" as though 15 billion years have passed since the "Big Bang" expansion process started.

When you accelerate a spaceship (or any object) along a certain axis of travel, you are changing the Permittivity and Permeability values within the atoms (mass) of which the spaceship is composed. If energy (E) = mc²; and c² (speed of light²) = $\frac{1}{\varepsilon_0\mu_0}$; then m (mass) = $\frac{E}{\varepsilon_0\mu_0}$.

The change in the magnitude of ($\varepsilon_0\mu_0$) within the atoms of the spacecraft produces the "appearance" of relative motion along a certain axis relative to other objects and observers. This is what causes the spaceship to "appear" to us as if it is moving in a certain direction. This is also what causes the inertial mass (weight) of the spaceship, which is measured in terms of acceleration (energy) -- (E = mc²) to increase as its relative velocity increases (see pages 71, 72, and 73).

What this means for us is that our perceptions of time, distance, relative motion, direction, gravity, energy, and mass are all being produced by various distortions in the Permittivity and Permeability values which represent the "effects" of the actual physical substance which is driving the expansion of our universe.

When Sam's spaceship accelerates toward the star that is 10 light years away, the energy which is used to accelerate his spaceship "distorts" or "reduces" the relative Permittivity and Permeability values within the atoms, and surrounding the atoms, of the spaceship, so that the spaceship begins to "appear" to us as though it has relative velocity in a certain direction, that time on board the spaceship slows down (relative to the Earth), and that our universe, from Sam's point of view, becomes less and less spatially extended.

The inertial mass of the spaceship will increase due to the fact that the electromagnetic waves of which the atoms in the spaceship consist, will be experiencing a greater and greater universal-expansion-related "rate of change" (velocity-related space-time distortion), which is also represented by the Permittivity and Permeability values.

The more the spaceship is accelerated, the greater this "rate of change" distortion becomes, in the expansion-related pressure reduction, which the physical substance has been undergoing during its entire expansion process since it started (see page 42). This expansion-related "rate of change" is referred to as "velocity-related" space-time distortion in this theory because that is the type of space-time distortion which produces the "effects" of relativity, including acceleration, energy, relative motion, gravitational fields, and relativistic mass increase (more on this in chapter 3).

Because of this relativistic mass increase, no massive object can ever achieve the speed of light (locally) no matter how much acceleration is applied. The speed of light is the maximum rate at which electromagnetic waves "appear to us" to propagate through our four-dimensional space-time continuum, so it represents the absolute speed limit in our universe, with the exception of our observations of the outer reaches of our universe, quantum teleportation, and quantum entanglements (more on this in chapter 5).

The speed of light represents the maximum amount of relative motion which any material object can attain (locally) because the Permittivity and Permeability values would be totally "distorted", or totally "reduced" within the atoms of that particular object.

Now just suppose for a moment that Sam's spaceship was actually able to attain the speed of light. The Permittivity and Permeability values in the c^2 portion (see page 81) of the $E = mc^2$ equation (mass of the spaceship) and also in the time and length equations, would be totally distorted (reduced to zero) and, consequently, Sam's perceptions of time and distance would no longer exist. As a result, the spaceship, with Sam on board, would no longer continue to experience the passage of time, nor continue to be spatially extended (size).

From Sam's point of view (observations), our whole universe, including the distance to the destination, and also the spaceship itself, would no longer continue to be spatially extended in the direction of travel, or in any other direction, for that matter. In other words, distance or size, as we know it, would be contracted down to (nearly) a single point in space, and time onboard the spaceship would completely stop. This means that one second of time onboard the spaceship would become infinitely long in comparison to one second of time back here on Earth.

Since light waves (their mass is practically zero) always travel, or propagate, "at the speed of light", then they are continually experiencing the maximum effects of space-time distortion, and thus, have a point-particle existence (photon), in addition to their electromagnetic wave existence.

This is also why the much higher-energy (frequency), and therefore, much more massive, electromagnetic wave/particles of which atoms consist, can exist at relative velocities which are less than the speed of light (more on this in chapter 4).

The theory of relativity is able to accurately describe the physical process by which these high-energy electromagnetic waves are contracted into point-particles. However, it soon became apparent that another theory was going to be needed which would be able to provide us with an accurate description of the fundamental nature, internal makeup, and the interactions which take place between other electromagnetic wave/particles to form atoms and molecules.

This led to the development of "Quantum Theory" and "Quantum Mechanics" which are explained in greater detail in Chapter 5. As more and more sophisticated experiments were designed to better-understand the effects of relativity and the atomic relationships of quantum physics, it soon became obvious to many physicists that the "Theory of Relativity" and "Quantum

Theory" were just two of the key components in the larger scheme of things.

One of the most prominent developments in this regard was that the continuing research in the field of Quantum Physics enabled us to gain further insight into the "fixed" and "quantized" states of electromagnetic wave/particles and their "probabilities" of interaction, which were frequently at odds with the dynamics of relativity. Thus, even more convincing evidence began to accumulate that there must be a deeper, underlying physical process going on which was simultaneously producing both the dynamic "effects" of relativity and also the "fixed" atomic states and probabilities of quantum physics.

The next step toward a better understanding of our relativistic, four-dimensional space-time energy field, which represents the "effects" of the mysterious "physical substance" which is driving the expansion of our universe, is to try to better-understand the nature of the realm in which our universe exists.

Since the relativity equations tell us that our perceptions of time and space are being produced by the expansion process of the mysterious "physical substance", then the void outside of our universe, beyond the maximum extent of the expansion of the "physical substance", could be considered to be completely empty space. In subsequent publications, this realm has been referred to as the "empty black void", but for the purposes of this theory, we shall refer to it as "basic reality" because the way that our universe would look to an observer in this realm represents what is "real" in our universe. First of all, an observer in this void, or realm of "basic reality", would not be perceptionally influenced by the huge, relativistic hologram (our space-time continuum) which represents our observable universe. Many of the strange characteristics of light waves, atoms, relativity paradoxes, and a variety of unusual quantum effects are being caused by the way in which our universe exists within this realm. Our perceptions of time and space, that we are accustomed to, would be meaningless in this realm. In other words, time, as we perceive it, would be completely stopped (become infinitely long), and space, or size, would be contracted down into a very small size (10^{-15} centimeters).

Now let's take a look at the "Time Chart" on page 42. The point of "basic reality", lower left, represents the point at which time and space are fully distorted according to the Relativity Theory. This point also represents the point from which our universe expanded according to the "Big Bang Theory". The main reason that the realm of "basic reality" is represented by a single point on the Time Chart is because that is how our universe would appear to an observer from the standpoint of "basic reality". . . as a single point or singularity. The "time line" represents the passage of time (for us) since our universe originated. The vertical line from the point of "basic reality", up to the start of the "time line", represents the "amount of concentration" or "pressure" changes in the physical substance, which can be measured in terms of acceleration according to the graphs on page 15.

(Continued on page 43)

TIME CHART

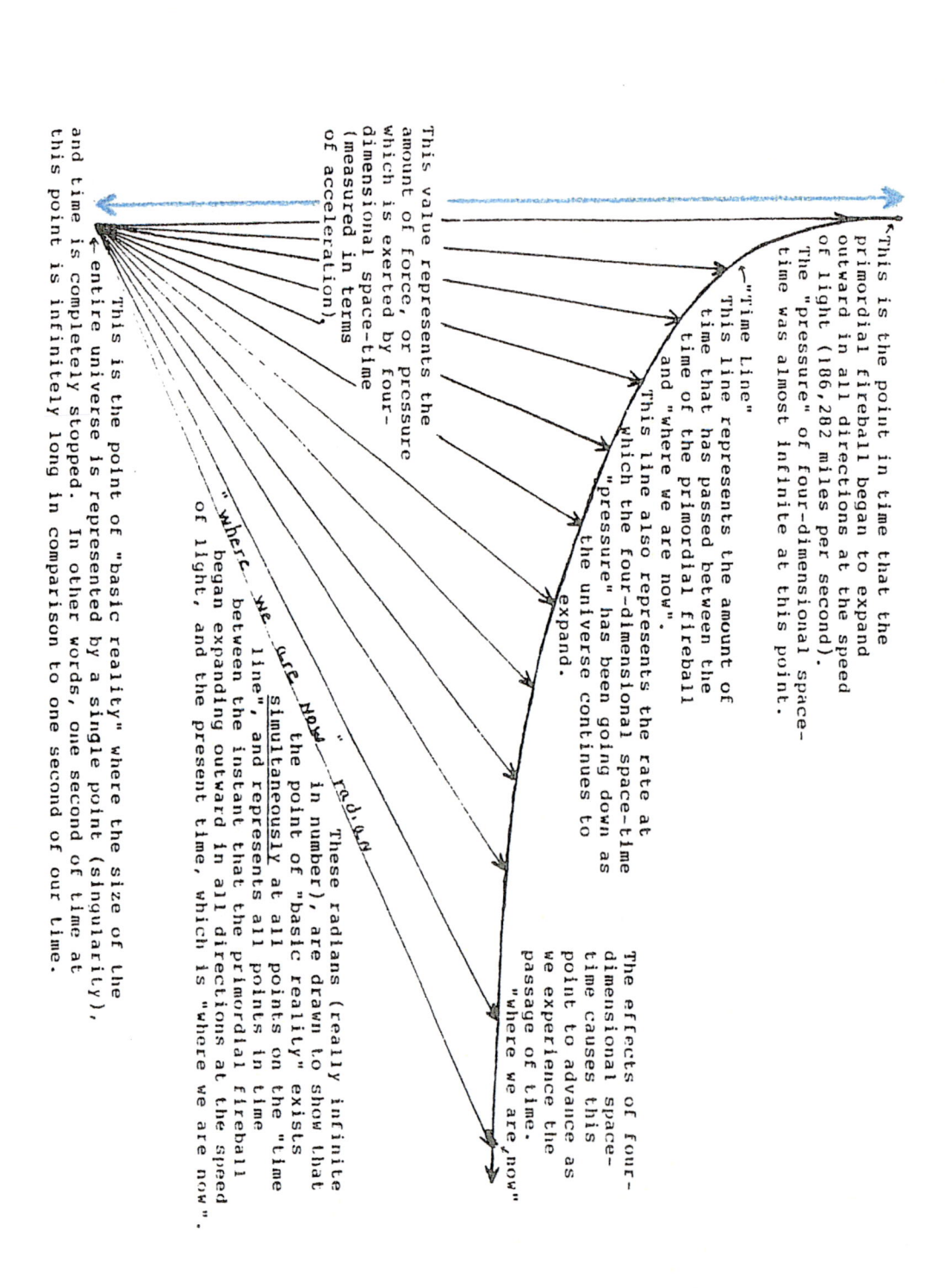

This is the point in time that the primordial fireball began to expand outward in all directions at the speed of light (186,282 miles per second). The "pressure" of four-dimensional space-time was almost infinite at this point.

"Time Line"

This line represents the amount of time that has passed between the time of the primordial fireball and "where we are now".

This line also represents the rate at which the four-dimensional space-time "pressure" has been going down as the universe continues to expand.

This value represents the amount of force, or pressure which is exerted by four-dimensional space-time (measured in terms of acceleration).

" rad. 0.0 "   we are now

These radians (really infinite in number), are drawn to show that the point of "basic reality" exists simultaneously at all points on the "time line", and represents the instant that the primordial fireball began expanding outward in all directions at the speed of light, and the present time, which is "where we are now".

The effects of four-dimensional space-time causes this point to advance as we experience the passage of time. "where we are now"

This is the point of "basic reality" where the size of the entire universe is represented by a single point (singularity), and time is completely stopped. In other words, one second of time at this point is infinitely long in comparison to one second of our time.

(Continued from page 41)

Changes in the "pressure" of the physical substance, which are directly related to acceleration, are what controls the rate at which all of the physical processes in our universe take place. In other words, changes in the "pressure" of the physical substance determine the relative strength of gravitational fields (acceleration), and consequently, how quickly events take place within our universe. Even though references to the physical presence and "pressure" of the physical substance are made in this chapter, the actual physical properties of the underlying physical substance have no real existence within our relativistic universe (as we see it) except, perhaps, within the atoms (more on this in chapter 5).

Ever since prehistoric times, we have always measured the passage of time by the rate at which physical processes such as the Earth's rotation, and orbit around the Sun take place. Since these processes are obviously controlled by the "pressure" of the physical substance, then it is the constantly changing "pressure" or concentration of the physical substance, as our universe continues to expand, which is really making time pass as we understand it.

Another way to look at this relationship is to consider Sam's spaceship as he is leaving Earth. If Bill uses a telescope to watch Sam's spaceship as he is departing, he would see Sam's spaceship getting smaller at a faster and faster rate as it accelerates away from Earth.

If Sam decided to stop accelerating his spaceship, then the spaceship would appear to be getting smaller and smaller at a steady rate which represents a constant velocity. Now if Sam decided to slow his spaceship down and come to a complete stop relative to the Earth, the spaceship, as it would appear to Bill, would stop getting smaller and stay the same size. This is how the Permittivity and Permeability values, which produce our perceptions regarding our four-dimensional space-time continuum, cause us to perceive mass, acceleration, distance, energy, and relative velocity.

ε_0 = Permittivity of free space = 8.854,187,817,6E-12 Farads per meter

μ_0 = Permeability of free space = 1.256,637,061,44E-7 Henry's per meter^{-1} (The $^{-1}$ exponent means that there is an inverse relationship between ε_0 and μ_0)

In regard to relative motion, it is not as if an object, such as Sam's spaceship, is moving through our four-dimensional space-time energy field per se. Instead, it is the velocity-related "rate of change" distortions, which the Permittivity and Permeability values represent, which are producing our perceptions regarding the relative motions of Sam's spaceship.

Now let's take a closer look at how the rate at which the "physical substance" is spreading out into empty space determines how we perceive distance. When we look across a room in our house and see objects on the far wall, we are really looking backward in time a small fraction of a second. The small fraction of a second that it took the light waves in the four-dimensional space-time energy field, which were reflected from those objects to reach our eyes, really represents a small difference in time between us and the objects at which we are looking.

In other words, the greater the distance to the object at which you are looking, the greater the time difference between you and that object, and consequently, the further back in time you are looking. This is why objects which are far away appear smaller than objects which are nearby.

Now, if you hang a mirror on the wall beside the object you are observing and then place another mirror on the wall behind the place where you are standing, you will find that you can align the mirrors directly at each other, so that you will see what looks like a whole string of mirrors stretching to infinity. You will also see several different images of yourself, each one smaller than the one in front.

According to this theory, this is caused by the constant expansion or "spreading out" of the physical substance, and the continual "stretching" of our four-dimensional space-time continuum, which causes time to pass as we perceive it. If the "pressure" or concentration of the physical substance were not constantly changing, the images in the mirrors would all appear to be the same size. In other words, as the light is being reflected from mirror to mirror, it is the space and time differences between each mirror that causes the images to appear smaller and, consequently, further away.

Eventually, if you could see far enough, the images in the mirrors would appear to be contracted down into a single point in space. This effect is the same as that which astronomers experience when they look out into our universe through a powerful telescope. The farther they look, the smaller the objects appear to be, and consequently, their telescope will need to have greater and greater levels of magnification for them to be able to see the more distant objects.

The further our astronomers are able see toward the outer reaches of our universe, the further backward in time toward the "big bang event" they are also looking. Consequently, we should exercise caution when interpreting our "most distant" observations due to the extreme violence and turbulence (in the space-time context) in the vicinity of the "big bang event" as our early universe was forming.

The "Space Chart" on page 20 shows us that the space relationship of our four-dimensional space-time energy field is like a huge sphere, with the point of "basic reality" on the "Time Chart" as its center point. The definition of a sphere is: The set of all points in space a certain distance from the center point.

If we could draw it, the "Space Chart" would also require an infinite number of radians to connect the center point (point of "basic reality") with the infinite number of points on the surface of the sphere. As with the "Time Chart", the center point of the sphere exists at all points on its surface simultaneously.

The size, or diameter, of the sphere represents the size of our universe as it has expanded and spread out from the point of the primordial singularity, until it reached the size it is today. The point of "basic reality", at the center of the sphere, also represents every point within, and on the surface of the sphere, simultaneously, when you superimpose the two charts (quantum theory). This will be explained in greater detail in chapter 5.

Many readers are quick to point out that this theory contradicts itself by stating that our universe still exists as a single point within the realm of "basic reality" in one paragraph, and then refers to the fact that our universe has expanded in a later paragraph. This dual existence of our universe, which causes electromagnetic waves to "appear" to us as though they have a dual existence, is caused by the strange nature of the realm of "basic reality" which will be explained in greater detail in Chapter 3. The observational inconsistencies, which are being caused by the nature of the realm of "basic reality", have produced several paradoxes in relation to the theory of relativity, in addition to various mathematical difficulties regarding attempts to define it.

Ironically, it was discovered later on that the mathematical relationships which define and unify the underlying physical processes which produce our perceptions and observations about our universe, are actually quite straightforward and simple. Perhaps many of our physicists have been using mathematics of a much higher level of complexity than that which is actually needed to produce a unification theory. This may also offer a possible explanation as to why a comprehensive unification theory has eluded us for so many years. In reality, it is the elementary patterns and repetitions in the mathematical formulas which have been most productive in our search for a unified field theory.

The information presented here is only a very rough outline of a total unification theory. There are many more details remaining to be discovered. The reader should be continually observant for additional numerical patterns and fundamental mathematical relationships which will continue to tie-together a very large number of diverse physical, chemical, and biological processes into the basic framework which is presented here.

The information that was set forth in this chapter helps us to lay the groundwork for a better understanding of the underlying physical processes which produce the effects of relativity and the forces which enable our universe to function according to our observations. A thorough understanding of these relationships will also help give us a much clearer description of how the information in the next few chapters will all fit together.

CHAPTER 3

GRAVITATIONAL, ELECTRIC, AND MAGNETIC FIELDS

Understanding the physical processes which produce the effects of relativity may require a slightly different way of looking at things, rather than the way we are accustomed to. Some of these concepts may challenge our previous understandings about time and space and, more generally, the world in which we live. However, every effort has been made to present this information in such a way that it will make this a very interesting subject for everyone to enjoy.

In the last chapter, we have seen how gravity, relative velocity and acceleration are able to change, or distort, the rate at which time proceeds. Since space and time are closely related, this change, or distortion, is usually concentrated in a particular area of space. The area of the greatest amount of change (space-time distortion) is usually at the center (four-dimensional view) of this area of space-time distortion.

When describing a localized area of space-time distortion, it is sometimes more convenient to imagine that four-dimensional space-time is represented by a flat sheet, so that this region of four-dimensional space-time distortion can now be visualized as a "dimple", or "distortion", in this flat sheet (two-dimensional view). This "dimple", or "distortion", has become known to physicists as a "manifold" in four-dimensional space-time (see page 51).

However convenient it might be to use a two-dimensional description, we should always keep in mind that this is actually a four-dimensional concept, three dimensions in space and one in time. The two dimensional view of a "manifold" in four-dimensional space-time is especially convenient when used to describe localized areas of space-time distortion.

Since we have already seen how acceleration can alter, or change the Permittivity and Permeability values in the c^2 portion of the $E = mc^2$ equation for mass, and also in the time and length equations which produce our perceptions of relative motion, now we shall take a look at a situation where changes in the Permittivity and Permeability values, in the space surrounding a massive object, produces acceleration. This phenomenon is known as gravity.

(Continued on page 52)

MANIFOLDS IN FOUR-DIMENSIONAL SPACE-TIME

This is the type of manifold (in the end of the "time line") which could be used to describe the earth's gravitational field (two-dimensional view).

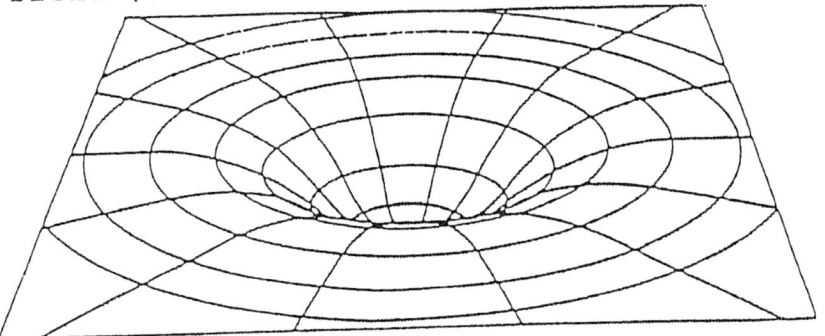

This type of manifold can also be used to describe a positive electric charge (above the "time line"), or a negative electric charge (below the "time line") at the point of "basic reality".

This is the type of manifold in four-dimensional space-time which would be created by a "black hole"

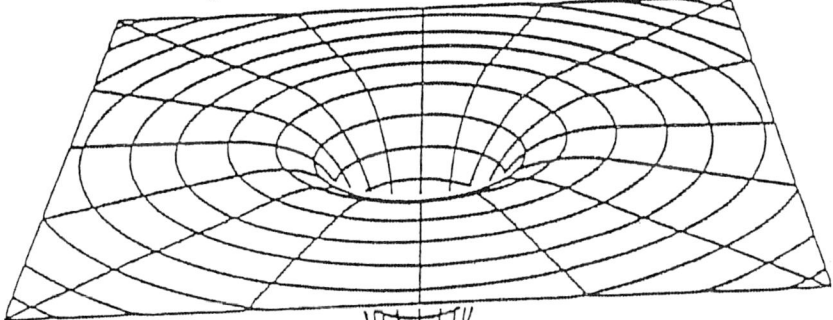

This manifold would extend back through the "time line" and into the primordial fireball singularity.

These manifolds actually represent a sphere, with the greatest amount of distortion at the center.

(Continued from page 50)

Experiments have shown that at the surface of the Earth, time is slowed down at a rate of approximately one second per thousand years.   This is a result of the Permittivity and Permeability values being "distorted" or "reduced" slightly due to the tremendous amount of relativistic wave motion within the large number of atoms in the Earth (more on this in Chapter 5).

ε_0 the Permittivity of free space: 0.000,000,000,008,854,187,817,6 Farads per meter

μ_0 the Permeability of free space: 0.000,000,125,663,706,144 Henry's per meter^{-1} (The $^{-1}$ exponent means that there is an inverse relationship between ε_0 and μ_0)

As we have seen before, relative velocity, acceleration, space, time, mass, gravity, and energy are all very closely related, as shown on the "Time Chart" on page 42, (see also page 71). Whenever you are considering velocity-related (expansion-related) space-time distortion, movement along "time line" from "where we are now" depends upon the amount of acceleration multiplied by the length of time that the acceleration was applied, to enable you to calculate the total relative velocity (see also the graphs on page 15).

When considering gravity at the surface of the Earth, as long as you are standing on the ground you are continuously being accelerated toward the center of the Earth, but you have no increase in relative velocity because the ground keeps you stationary. In other words, you are experiencing the effects of velocity-related space-time distortion which places you back a certain distance along the "time line" (see page 42), but you are not moving continually back along the "time line" as you would be if you had jumped out of an airplane and your relative velocity was steadily increasing as you are falling faster and faster (being continuously accelerated) toward the surface of the Earth.

Another important point to remember here is that it is sometimes much simpler and easier to consider just the time dilation effects when calculating the relative strength of gravitational fields. However, this is a space-time phenomenon, so you really have to consider both the time dilation effects in addition to the length (size) contraction effects when you are working with gravitational fields.

The best way to look at this relationship on the "Time Chart" (page 42) is to imagine that your location on the "time line", which represents your reference frame (surface of the Earth), is not quite as close to the end of the arrow as it would be if you were out in space far away from any sizeable gravitational fields.

When you are standing on the surface of the Earth, the acceleration which you are experiencing is a direct result of the difference (T – T' = 9.8 meters/second^2) between the rate of the passage of time far out in space and the rate of the passage of time on the surface of the Earth. This acceleration rate of 9.8 meters/second^2 corresponds to a measured reduction in the rate of the passage of time which is equal to approximately one second per thousand years.

This direct relationship between the rate of reduction in the passage of time (the reduction in the space-time coordinates) and the corresponding increasing acceleration, can be mathematically calculated, precisely, at all points on the relativity curve (see page 71) --- all the way back to the primordial singularity. According to the "Time chart" on page 42, as the strength of the gravitational field (acceleration) increases, then your location on the "time line" is moved further and further back from the arrow point.

Let's take a look at the "Time Chart" on page 42 again, and turn it so the arrow at the end of the "time line" is pointing straight at you. Next, imagine that the end of the arrow was cut off – revealing a cross section of the end of the "time line".

Now, imagine that the end of the "time line" (cross section) that you are looking at is round, and could be magnified to cover an area the diameter of our universe. You are now prepared to visualize a two-dimensional view of the type of four-dimensional space-time distortion which is created in a gravitational field (see page 54). Actually, the graphs and drawings on pages 15, 42, 51, 55, 68, and 71, are all slightly different representations of our four-dimensional space-time energy field (Permittivity and Permeability values) which produce our perceptions of our four-dimensional space-time continuum.

(Continued on page 56)

GRAVITATIONAL MANIFOLD TIME CHART

(as it can be used to represent a gravitational field).

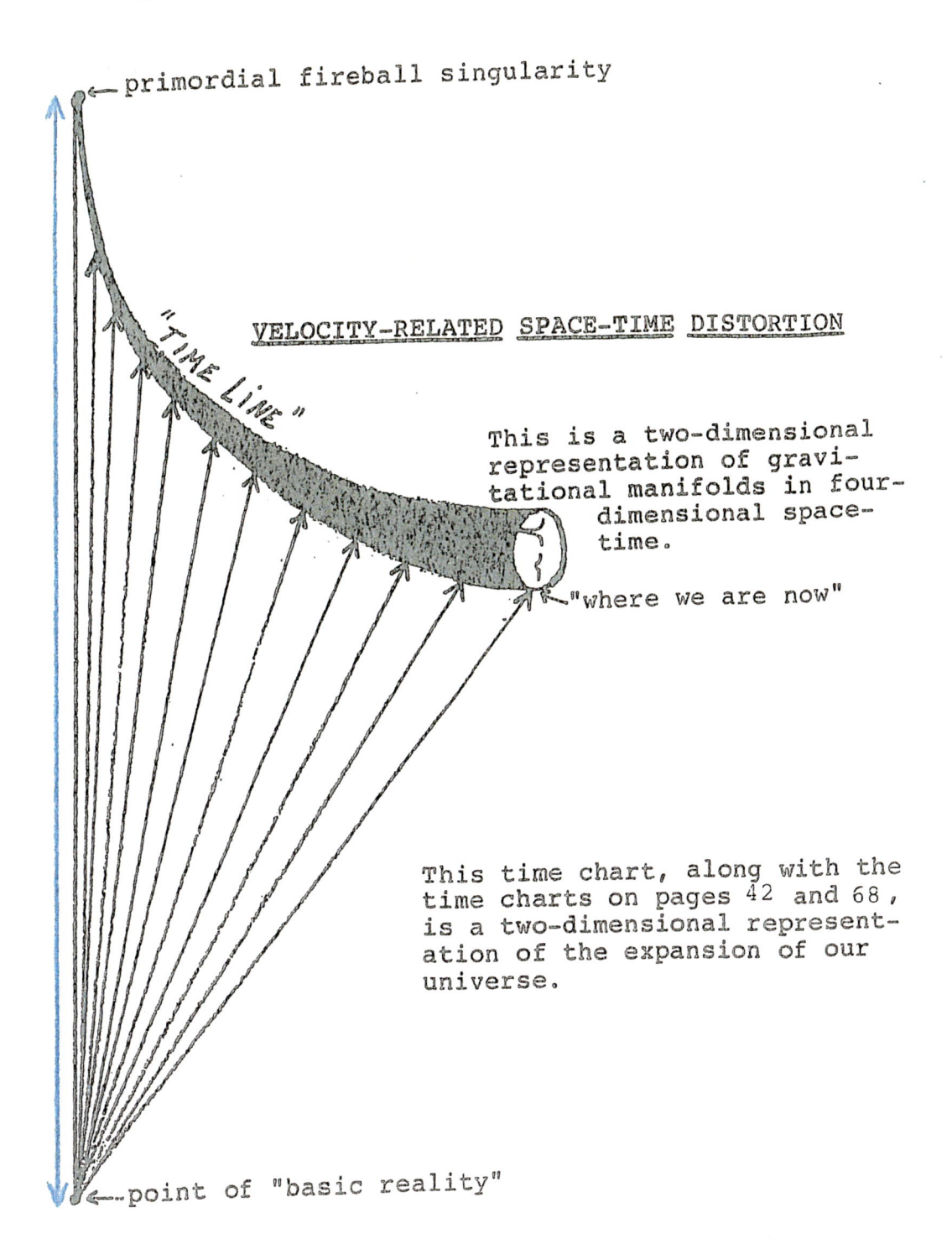

← primordial fireball singularity

"TIME LINE"

VELOCITY-RELATED SPACE-TIME DISTORTION

This is a two-dimensional representation of gravitational manifolds in four-dimensional space-time.

← "where we are now"

This time chart, along with the time charts on pages 42 and 68, is a two-dimensional representation of the expansion of our universe.

← point of "basic reality"

55

(Continued from page 54)

We can now consider a gravitational field to be a "dimple" in this circle which represents an end-view of the "time line".

As the strength of the gravitational field (acceleration) is increased, imagine the bottom of the dimple (manifold) going deeper into the "time line", back toward the primordial singularity.

The manifold which represents the reduction in the Permittivity and Permeability values within the gravitational field of the Earth, would look like only a small distortion in the end of the "time line" in comparison to that of a "black hole" which would extend all the way back through the "time line" and into the primordial singularity.

When a star, larger than two and a half times the mass of our Sun burns out, most of the lighter elements in the star are blown away in a tremendous super-nova explosion, leaving most of the remaining heavier elements in an extremely dense, massive core which is just one of the possible outcomes.

The gravitational field (space-time distortion) surrounding this extremely massive core can be so great that all the atoms (within the core) can be crushed and compressed into a single super-neutron, or singularity where the space-time energy field is almost totally distorted. This has become known to physicists as a "Black Hole" in four-dimensional space-time.

If the strength of the Earth's gravitational field was increased to equal that of a "Black Hole", your location on the "time line" would be moved all the way back to the starting point in time when the primordial singularity first began to expand. At that point, the acceleration produced by the gravitational field would be so powerful that you, and the Earth too, would be crushed and compressed down to the size of a neutron (1.65×10^{-13} centimeter in diameter).

This is an awesome testimony to the tremendous pressure difference of the "physical substance" inside the primordial singularity at the instant that it began to expand outward, and that which it is today (as it has expanded and spread out).

This "Black Hole" singularity, which actually represents a percentage of the original primordial singularity, now becomes the center of a region of space where the Permittivity and Permeability values are reduced to the point where time almost completely stops and distance, or size, is almost completely contracted down to a single point in space. According to this theory, however, *total* space-time distortion could be achieved by a black hole singularity only if the total mass of our whole universe was once again contained within this singularity.

Our mysterious space-time energy field, which is produced by the "effects" of the actual physical substance, has several other strange characteristics which we shall attempt to describe here. One of these is its ability to produce an electric charge.

(Continued on page 63)

POSITIVE ELECTRIC CHARGE

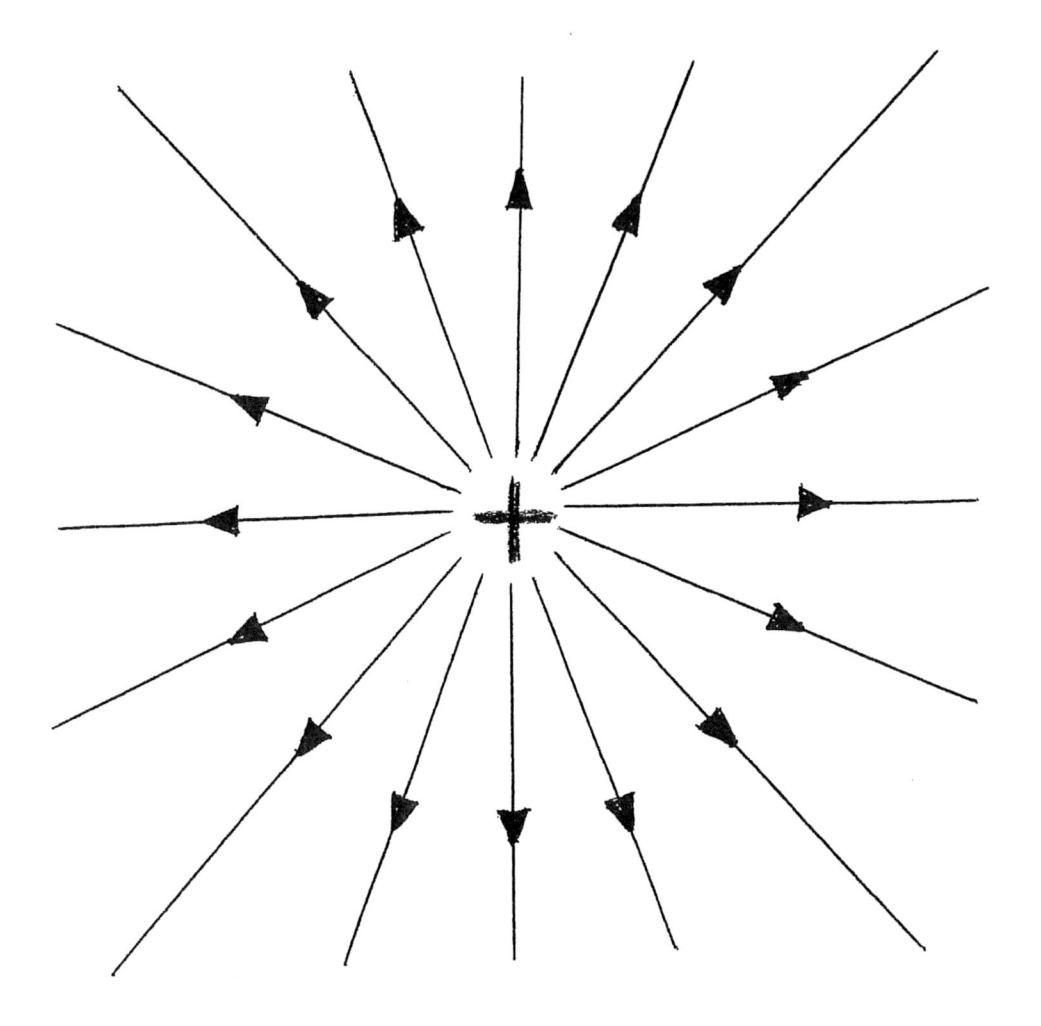

This drawing describes a region of space which is under the influence of a higher "pressure" (than the median "pressure") of the "physical substance" within the primordial singularity as it exists in the realm of "basic reality". The median "pressure" of the physical substance is represented by the "time line" on page 68. The positive electric charge is represented by a (green) manifold above the "time line".

NEGATIVE ELECTRIC CHARGE

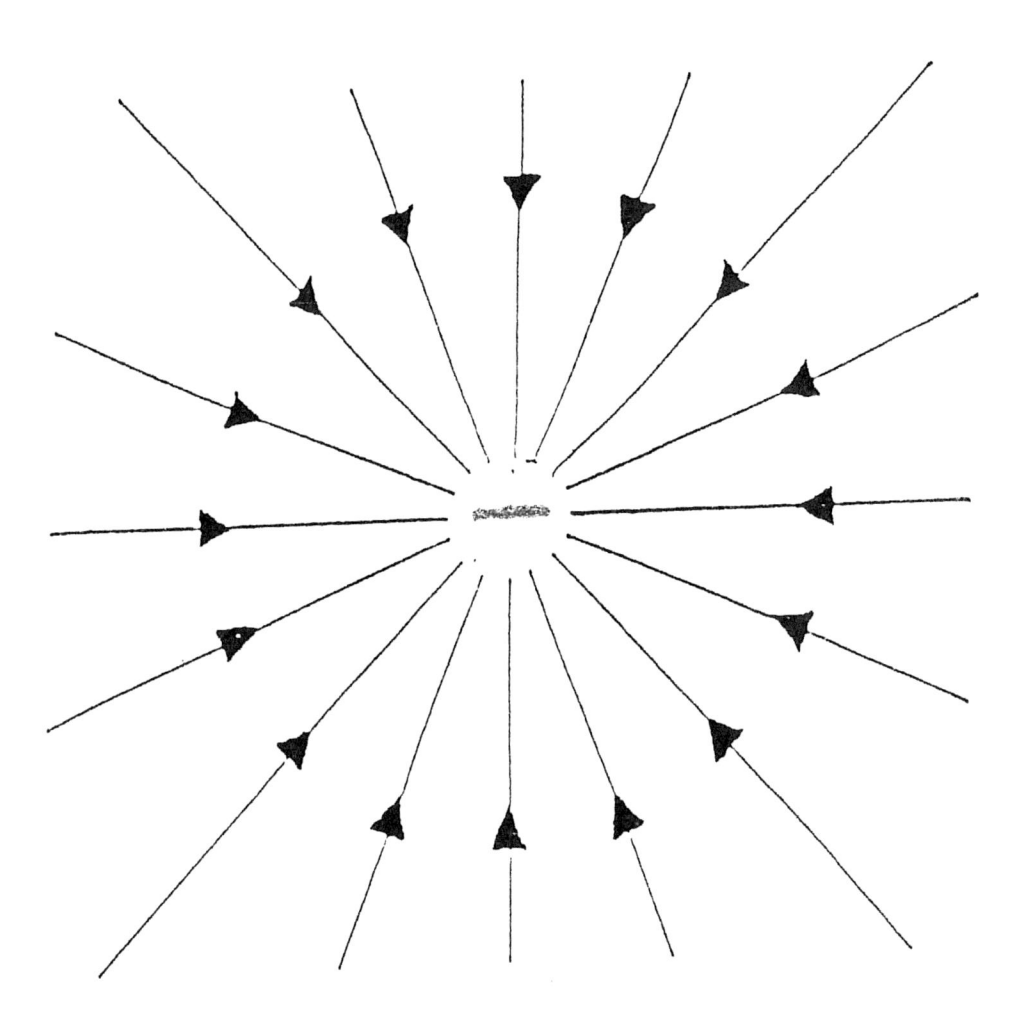

This drawing describes a region of space which is under the influence of a lower "pressure" (than the median "pressure") of the physical substance within the primordial singularity as it exists in the realm of "basic reality". The median "pressure" of the "physical substance" is represented by the "time line" on page 68. The negative electric charge is represented by a (red) manifold below the "time line".

TWO POSITIVE ELECTRIC CHARGES

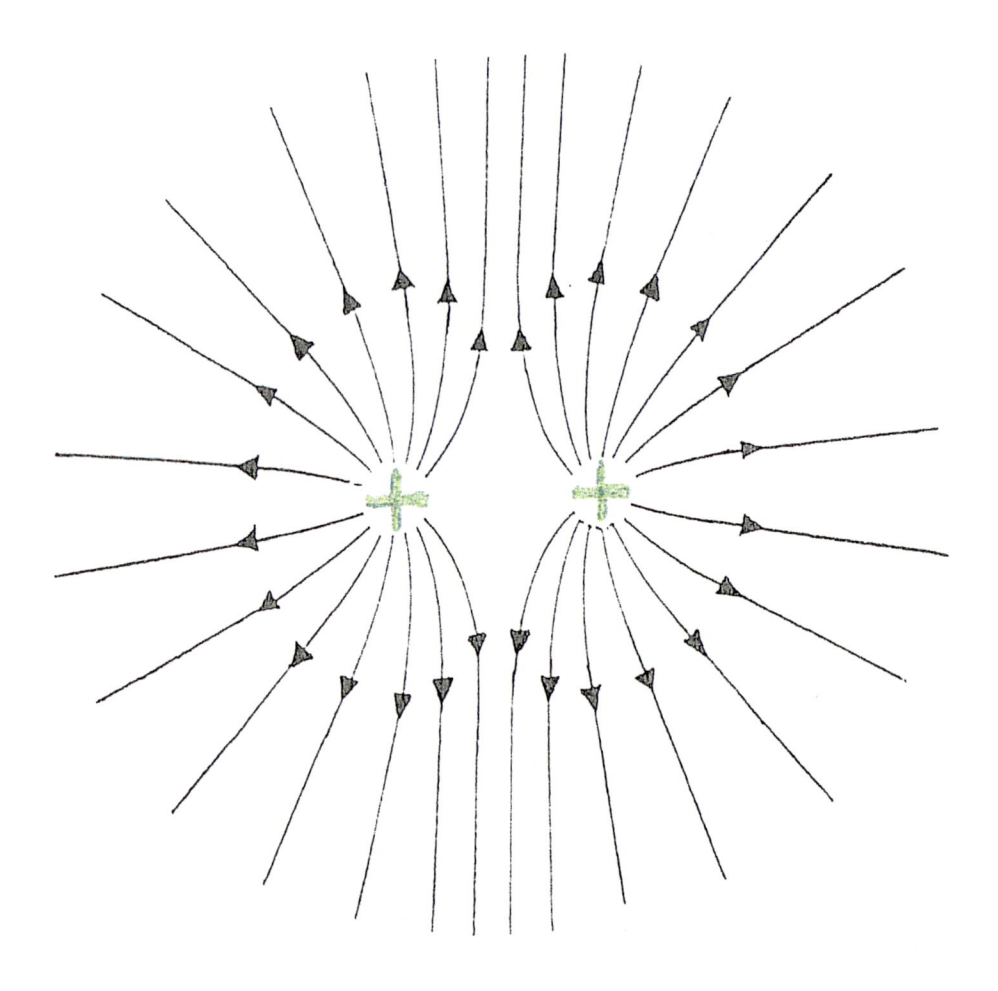

The two positive electrical charges shown here represent regions of space which are under the influence of a higher "pressure" (than the median "pressure") of the "physical substance" within the primordial singularity (see page 68). Since physical "pressure" differences in the "physical substance" always tend to seek the median "pressure" or equilibrium, as it exists at the point of "basic reality", then these two charges will repel each other in order to remain as close to equilibrium as possible.

TWO NEGATIVE ELECTRIC CHARGES

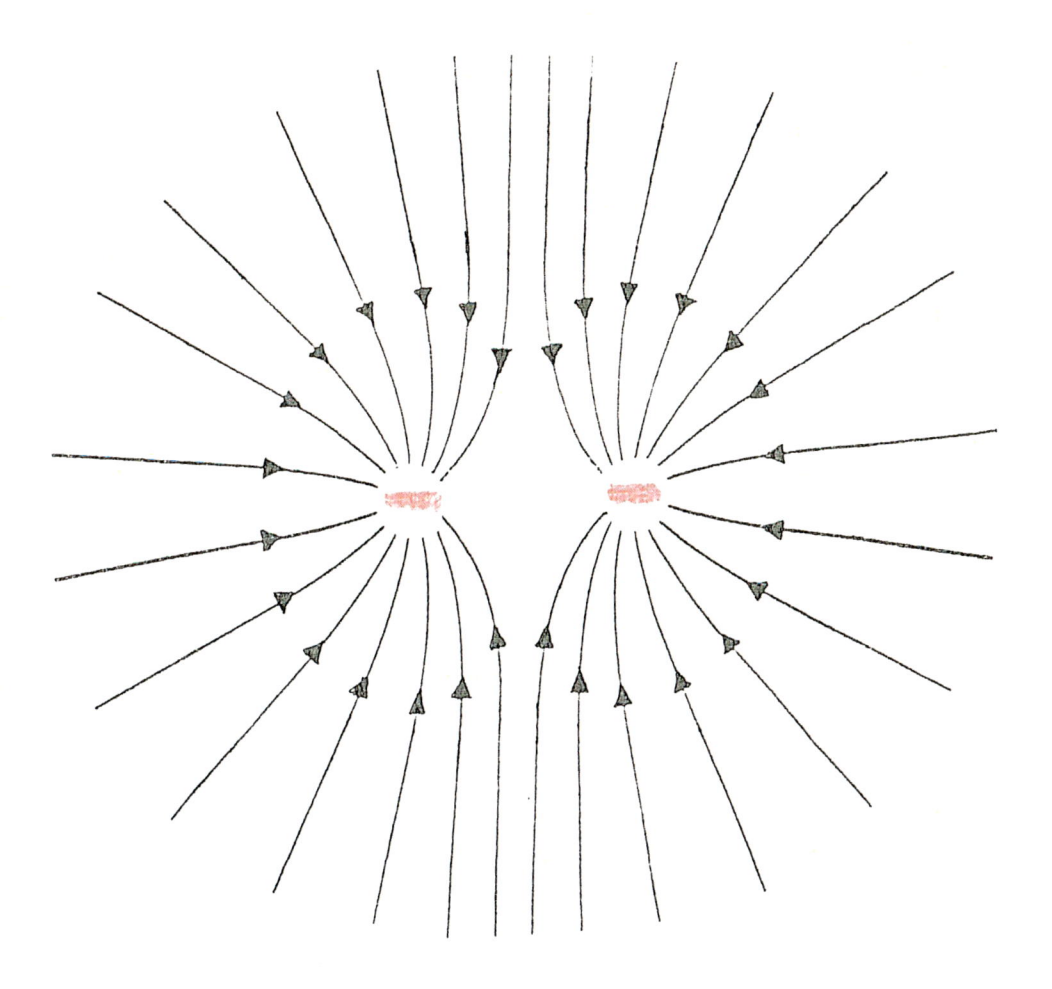

The two negative electrical charges shown here represent regions of space which are under the influence of a lower "pressure" (than the median "pressure") of the "physical substance" within the primordial singularity (see page 68). Since physical "pressure" differences in the "physical substance" always tend to seek the median "pressure" or equilibrium, as it exists at the point of "basic reality", then these two charges will repel each other in order to remain as close to equilibrium as possible.

OPPOSITE ELECTRIC CHARGES ATTRACT

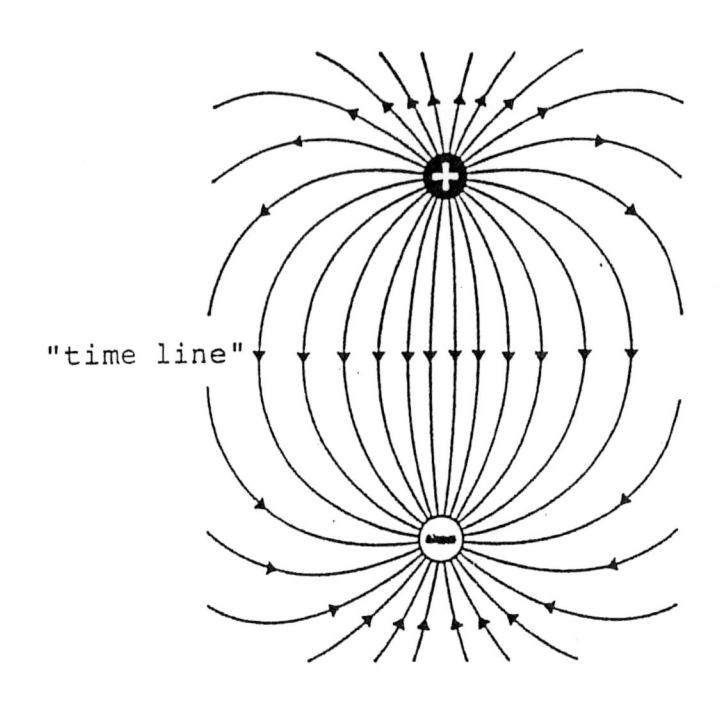

"time line"

    Opposite electrical charges tend to attract because they represent "pressure" differences in four-dimensional space-time. This is why they are shown on opposite sides of the "time line", which is represented by the horizontal line of arrows above.
    The acceleration that they produce is always toward equilibrium, or the median "pressure" of four-dimensional space-time as it exists at the point of "basic reality". The point of "basic reality" (in this case) represents the "real" existence of the four-dimensional space-time in our universe as we see it (our dimension).

(Continued from page 57)

To gain a better understanding of what an electric charge really is, we will need to return, once again, to the time and length equations which Albert Einstein used in his Special Theory of Relativity to describe our four-dimensional space-time continuum. In this particular instance, however, we will also be using the information on pages 30, 68, 71, and 81 for reference purposes.

On these pages, we can see that the rate at which light waves (electromagnetic waves) propagate through our space-time energy field are determined by two values, the Permittivity and Permeability of free space, which may also be considered to be similar to universal "spring constants" by comparison.

ε_0 the Permittivity of free space: 8.854,187,817,6E-12 Farads per meter.

μ_0 the Permeability of free space: 1.256,637,061,44E-7 Henry's per meter^{-1} (The $^{-1}$ exponent means that there is an inverse relationship between ε_0 and μ_0).

The Permittivity and Permeability values, taken together, represent the actual physical properties of our universal space-time energy field, and thus, they determine the effects of relativity, including the rate of propagation of electromagnetic waves, as well as the local strengths of gravitational fields, electric charges, and magnetic fields.

The Permittivity and Permeability values represent the "effects" of physical pressure and "rate of change" disturbances in the actual "physical substance" within the primordial singularity. We know this because electromagnetic waves are both "lensed" and "frequency-shifted" by gravitational fields, and that they also experience Doppler frequency shifts associated with relative motion either toward or away from the observer (see "Doppler shift" formula on page 72).

This strongly indicates that the Permittivity and Permeability of free space, are not really constants after all, and so, under certain circumstances, they also can become variable.

A positive electrical charge, then, may be considered to be a higher-pressure "distortion" in the actual "physical substance" which is represented by an increase in the Permittivity value by itself (see drawings on pages 58 and 68).

A negative electrical charge can be considered as a lower-pressure "distortion" in the actual "physical substance" which is represented by a decrease in the Permittivity value by itself (see drawings on pages 59 and 68) for a particular region of space.

Our perceptions of positive and negative electric charges are a direct result of the specific Permittivity value at all points in space where the positive or negative electric charges are located. In addition to this, physical pressure differences (positive and negative electric charges) in the "physical substance" are always perpendicular to expansion-related (velocity-related) distortions such as: gravity; relative motion; space; time;

mass; and energy; in the "physical substance" (see drawings on pages 55 and 68).

The expansion-related (velocity-related) distortions in the four-dimensional space-time energy field are a direct result of "equal" variations in both the Permittivity and Permeability values. Consequently, the difference in the Permittivity values in the space surrounding electric charges are responsible for the attraction (acceleration) of two opposite electric charges, and the repulsion (acceleration) of two like electrical charges (see pages 60, 61, 62, and 73). Also, the Permittivity and Permeability values are perpendicular to each other which has been demonstrated by laboratory experimentation regarding electromagnetic phenomena.

Referring back to the "Time Chart" on page 42, you can see that at the instant that the primordial singularity began to expand, the "pressure" of the physical substance was very high. Then, as the physical substance continued to expand and spread out into "empty" space, the pressure of the physical substance began going down in direct proportion to the rate at which it was expanding according to the graphs on page 15.

From the information on pages 15, 42, 71, and 81, it has become apparent that the Permittivity and Permeability values are direct representations of the actual physical conditions (pressure and "rate of change" distortions) within the "physical substance" which is driving the expansion of our universe.

By now you are probably wondering why Albert Michelson and Edward Morley could not directly detect the presence of the "physical substance" (ether) since that is what is "really" causing our universe to expand. This question can be answered by pointing out that our universe (as we perceive it) functions according to the Theory of Relativity which describes the "effects" of the "physical substance" (four-dimensional space-time distortion).

When considering the relative motions of the Earth, it now becomes obvious that the Earth is not actually moving through the "physical substance", or even through the four-dimensional space-time energy field, per se. Instead, distortions in the expansion process of the "physical substance" (velocity-related space-time distortion via the Permittivity and Permeability values) are what is actually producing our perceptions about the relative motions of the Earth.

We can now see that the way the Michelson-Morley experiment was set up, it is no wonder that it failed to detect the presence of the "physical substance". Since there were no gravitational differences, or no relative motion differences between the light source and the detector, (the whole unit was mounted on a solid platform), there could not possibly have been any phase shift between the two light beams which were being compared by the detector. This was simply the wrong type of experiment for detecting the actual presence of the "physical substance".

The "real" expansion of our universe as seen by an observer in the realm of "basic reality" is a combination of pressure-related, and "change" types of distortions within the "physical substance". Since the concepts of time and space, as we understand them, are meaningless within the realm of "basic reality", then this pressure-related expansion process of the "physical substance" cannot occur in an even and continuous manner. It actually occurs in "steps" so that every time the primordial singularity gets a little bit larger, it takes on a whole new existence (see page 136). These separate existences of our universe are referred to in this theory as "Dimensions" which should not be confused with the four dimensions of height, length, width, and time.

The reason that the "physical substance" has a real existence in our "Dimension" only at the point of "basic reality" (primordial singularity) is due to the fact that our "Dimension" came into existence very early in the expansion process while the primordial singularity was very small and highly concentrated. The passage of time and the other relativistic effects in our "Dimension" are caused by the "real" pressure-related expansion and spreading out of the "physical substance" as it is occurring in other "Dimensions" (existences), (see page 136).

Now imagine the "time line" as a wide piece of ribbon being seen from the end, as it is shown on page 68. From this point of view, a positive electrical charge can now be represented by a vertical manifold (Permittivity value increase) above the "time line", starting at the point of "where we are now" and extending all the way back to the primordial singularity.

(Continued on page 69)

TIME CHART WHICH REPRESENTS ELECTRIC CHARGES

TIME CHART
(as it can be used to represent electric charges)

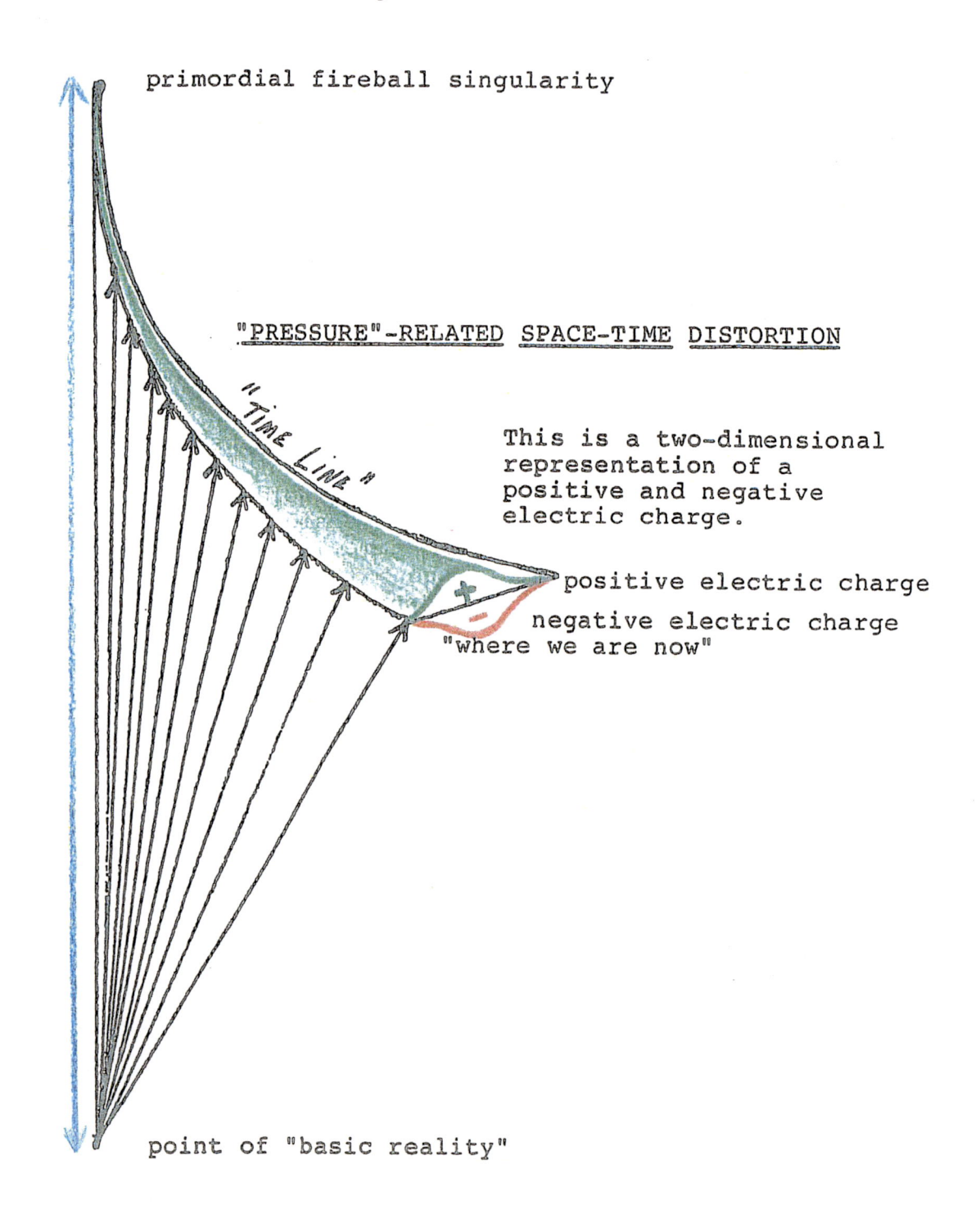

primordial fireball singularity

"PRESSURE"-RELATED SPACE-TIME DISTORTION

"TIME LINE"

This is a two-dimensional representation of a positive and negative electric charge.

positive electric charge

negative electric charge

"where we are now"

point of "basic reality"

(Continued from page 67)

A negative electric charge can now be visualized as a vertical manifold (Permittivity value reduction) below the "time line" starting at the point of "where we are now" and extending all the way back into the primordial singularity.

The reason that these manifolds are drawn in this manner is to show that the "effects" of physical distortions (electric and magnetic fields) in the "physical substance" are instantaneous across space and time between the primordial singularity (which still exists in the realm of "basic reality") and "where we are now".

The "time line" in this case, refers to the median pressure (Permittivity and Permeability values) of the "physical substance" at any point in time between the instant that the primordial singularity began expanding outward, and "where we are now".

However, since electric fields are physical distortions in the actual "physical substance", they only have a real existence at the point of "basic reality", which also represents the primordial singularity. We can only experience the "effects" of physical distortions in the actual "physical substance" in our time frame.

Experiments have shown that two electric charges that are alike will be accelerated away from each other (see drawings on page 60 and 61) while two electric charges which are opposite will be accelerated toward each other (see drawing on page 62).

The reason for this phenomenon is that the "time line" represents the point of equilibrium, or neutral, when considering the median pressure of the "physical substance", and consequently, the acceleration produced by, or between, electric charges is always toward equilibrium or neutral.

ε_0 the Permittivity of free space: 8.854,187,817,6E-12 Farads per meter

μ_0 the Permeability of free space: 1.256,637,061,44E-7 Henry's per meter^{-1} (The $^{-1}$ exponent means that there is an inverse relationship between ε_0 and μ_0)

Now when you hold two like electrical polarities (charges) together, their respective charges are added, so that you end up increasing the Permittivity value in that region of space -- above the equilibrium Permittivity value (see pages 60, 61, 68, and 73). Since the two electric charges are always trying to achieve equilibrium, then the two like charges will repel each other in order for that region of space to remain as close to equilibrium as possible.

The case is different with two opposite electrical charges. Since they exist on opposite sides of the "time line", and since the acceleration they produce is always toward the equilibrium Permittivity value, or neutral, then they are attracted toward each other and try to cancel each other out.

(Continued on page 74)

SPACE-TIME DISTORTION CURVE

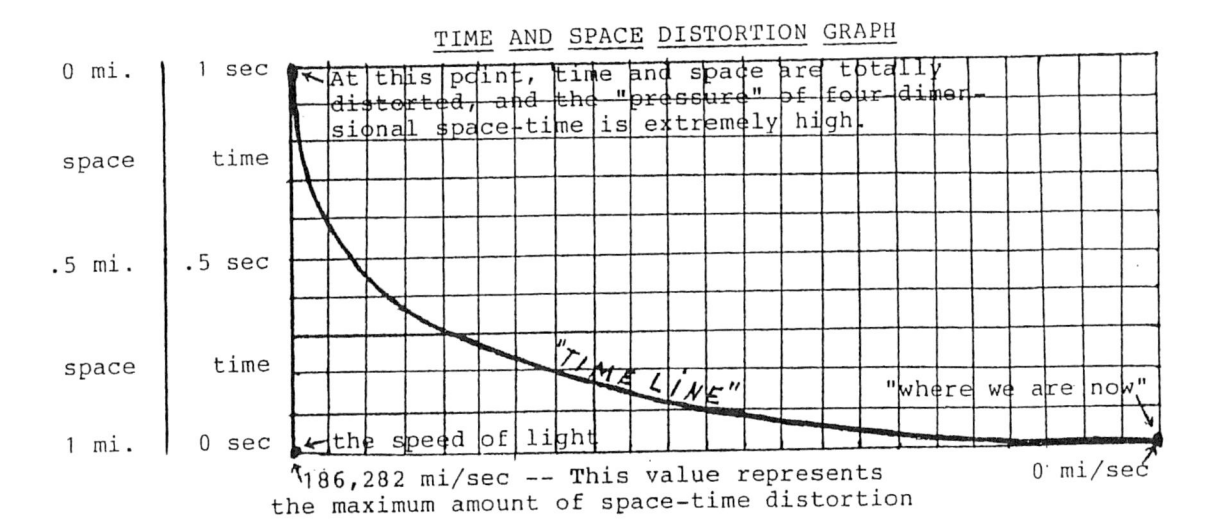

TIME AND SPACE DISTORTION GRAPH

At this point, time and space are totally distorted, and the "pressure" of four-dimensional space-time is extremely high.

"TIME LINE"

"where we are now"

← the speed of light

186,282 mi/sec -- This value represents the maximum amount of space-time distortion

0 mi/sec

The graphs of the "time" and "length" equations on page 15 reveal the underlying physical process which is producing our space-time continuum. Note that the form that all these equations take is the same. This indicates that our perceptions about our universe including: space; time; mass; gravity; relative motion, energy; and wave motion are all being produced by the same underlying physical process.

LENGTH (SPACE): This equation describes the rate at which length within an inertial reference frame decreases as relative motion increases. Gravitational fields are represented by the time and length equations together (see page 15).

$$L' = L\sqrt{1 - \frac{v^2}{c^2}} \quad ; \quad \text{also} \quad L' = L\sqrt{1 - \frac{(aT)^2}{\varepsilon_0 \mu_0}}$$

TIME: This equation describes the rate at which time within an inertial reference frame slows down (1 second becomes longer) as relative motion increases. Gravitational fields are represented by the time and length equations together (see page 15).

$$T' = \frac{T}{\sqrt{1 - \frac{v^2}{c^2}}} \quad ; \quad \text{also} \quad T' = \frac{T}{\sqrt{1 - \frac{(aT)^2}{\varepsilon_0 \mu_0}}}$$

MASS: This equation describes the rate at which mass increases within an inertial reference frame as its relative motion increases.

$$m = \frac{m_0}{\sqrt{1 - \frac{v^2}{c^2}}} \quad ; \quad \text{also} \quad m = \frac{m_0}{\sqrt{1 - \frac{(aT)^2}{\varepsilon_0 \mu_0}}}$$

RELATIVE MOTION: This equation describes the rate at which relative motion increases as acceleration is applied over a certain time period.

$$v = \frac{at}{\sqrt{1 - \frac{v^2}{c^2}}} \quad ; \quad \text{also} \quad v = \frac{at}{\sqrt{1 - \frac{(aT)^2}{\varepsilon_0 \mu_0}}}$$

WAVE MOTION: This equation describes the relationship between uniform circular motion and electromagnetic wave motion which is the most basic form of energy within our universe. For additional relationships between wave motion, mass, relative motion, kinetic energy, and potential energy (see page 88).

$$v = v_m \sqrt{1 - \frac{X^2}{A^2}} \quad ; \quad \text{also} \quad v = \frac{at}{\sqrt{1 - \frac{(aT)^2}{\varepsilon_0 \mu_0}}}$$

RELATIVISTIC DOPPLER EFFECT: This equation describes the change in frequency ($f - f'$), of electromagnetic waves (light) which are emitted by an object with velocity (relative motion) between the object and an observer. Theta (θ) is the angle between the direction in which the object is moving, and the direction of "line-of-sight" between the object and the observer.

$$f' = f \frac{1 - \frac{v}{c} \cos\theta}{\sqrt{1 - \frac{v^2}{c^2}}} \quad ; \quad \text{also} \quad f' = f \frac{1 - \frac{v}{c} \cos\theta}{\sqrt{1 - \frac{(aT)^2}{\varepsilon_0 \mu_0}}}$$

1/r² LAW: This term describes the rate at which the acceleration produced by electric, magnetic, and gravitational fields decreases with distance " r " (radius) from the object which is producing the field. When the terms below are solved for increasing values of " r ", the curve which is plotted on a graph is the same as the curve for all the equations above, thus proving that each of these forces are fundamentally related through a common source --- the Permittivity and Permeability values of free space (ε_0 and μ_0). Force (F) = mass (m) X acceleration (a); mass = $E(\varepsilon_0 \mu_0)$

GRAVITY: $\quad$ Force $= E(\varepsilon_0\mu_0)a = G\dfrac{m_1 m_2}{r^2}$

ELECTRIC CHARGE: $\quad$ Force $= E(\varepsilon_0\mu_0)a = k_e \dfrac{q_1 q_2}{r^2}$

MAGNETIC FIELD: $\quad$ Force $= E(\varepsilon_0\mu_0)a = 2k_A \dfrac{I_1 I_2}{r^2}$

LEGEND:

a = Acceleration
A = Maximum displacement from the equilibrium position (see page 88)
c = Speed of light in a vacuum
f = frequency
f' = Distorted frequency
E = Energy -- Newton · meters
F = Force -- in Newtons; 1 Newton = 1 kilogram · meter/second²
G = Gravitational constant: 6.673E-11 Newton · (meters/kilogram)²
I = direct current moving through the wire -- measured in Amperes
k_A = magnetic force constant: $2k_A = \mu_0/2\pi$ = 2E-7 Newtons/meter
k_e = Coulomb's constant: 8.987,551,787,368,176,4E-16 N · m²/C²
L = Length in "flat" space (away from any strong gravitational fields).
L' = Distorted length
m = mass -- measured in grams
m₀ = Original mass
q = electric charge -- measured in Coulomb's
r = Distance (radius) from the object (or between the objects) which is/are producing the field -- measured in meters
t = Length of time that the acceleration was applied.
T = Time in "flat" space (away from any strong gravitational fields).
T' = Distorted time
V = Velocity
V_m = Maximum Velocity
X = Displacement from the equilibrium position (see page 88).

(Continued from page 70)

Velocity-related space-time distortion always lies along the axis of the "time line", whereas the positive or negative space-time distortion created by an electric charge, is always perpendicular (above and below) the "time line".

The close relationship between electric, magnetic, and gravitational fields is also demonstrated by the $\frac{1}{R^2}$ law, better known as the inverse square law.

This mathematical relationship describes how the strength of a gravitational, electric, or magnetic field diminishes with distance. This is another strong indication that these different types of fields all have a common source.

The $\frac{1}{R^2}$ law plots the same curve on a graph as the relativity equations on page 15, but in the reverse direction. This is also another way that we can demonstrate that gravitational, electric, and magnetic fields are just different types of distortions (perpendicular) in the same "physical substance" (see pages 42, 51, 55, 68, and 71) which is represented by the Permittivity and Permeability values.

Magnetism is directly related to a movement, or "rate of change" distortion within the physical substance itself (see drawings on page 75 and 76).

(Continued on page 77)

HYDROGEN ATOM

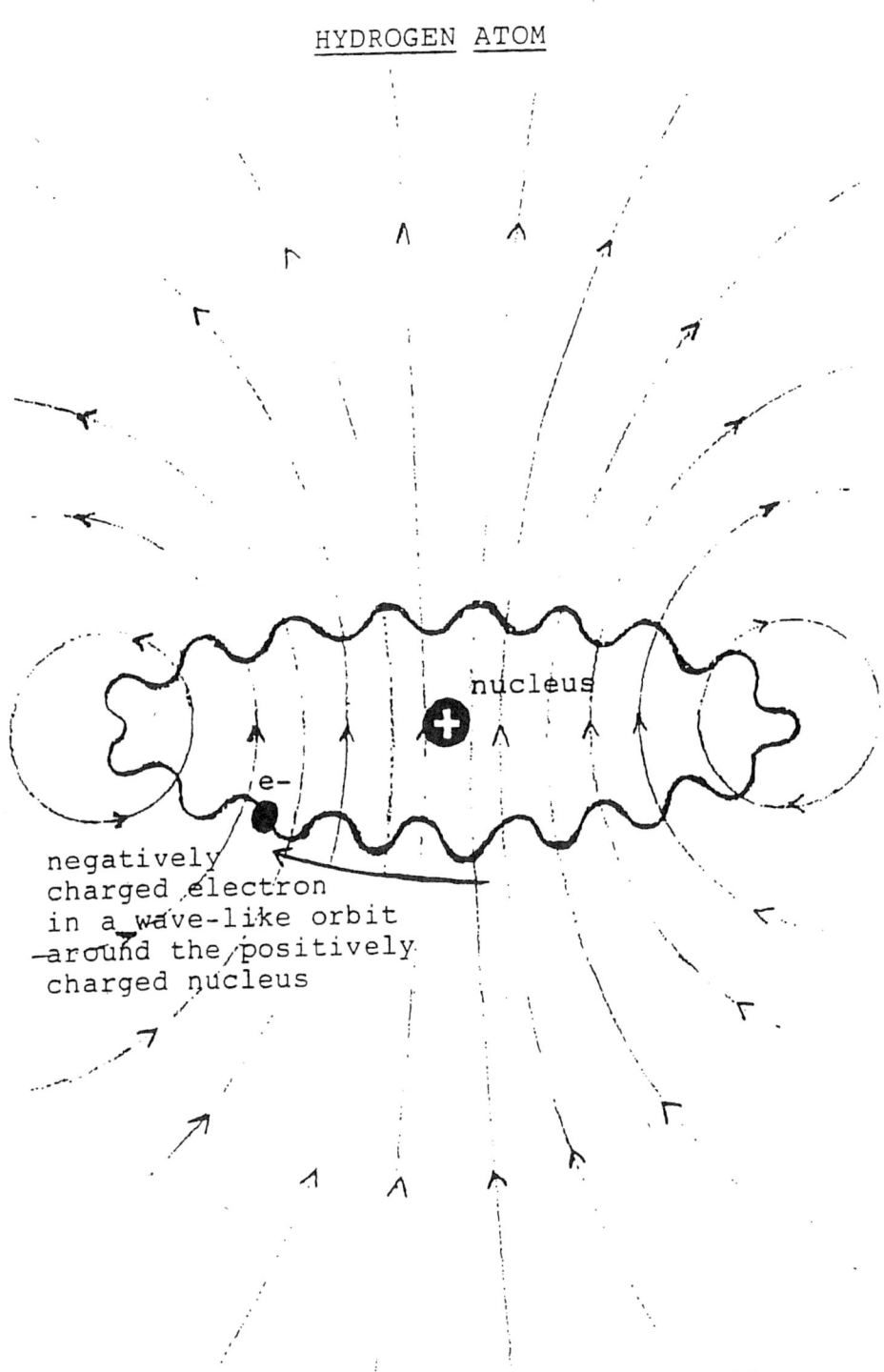

negatively
charged electron
in a wave-like orbit
around the positively
charged nucleus

these magnetic lines of flux represent the
movement of four-dimensional space-time
created by the moving electrical charge
(electron).

BAR MAGNET

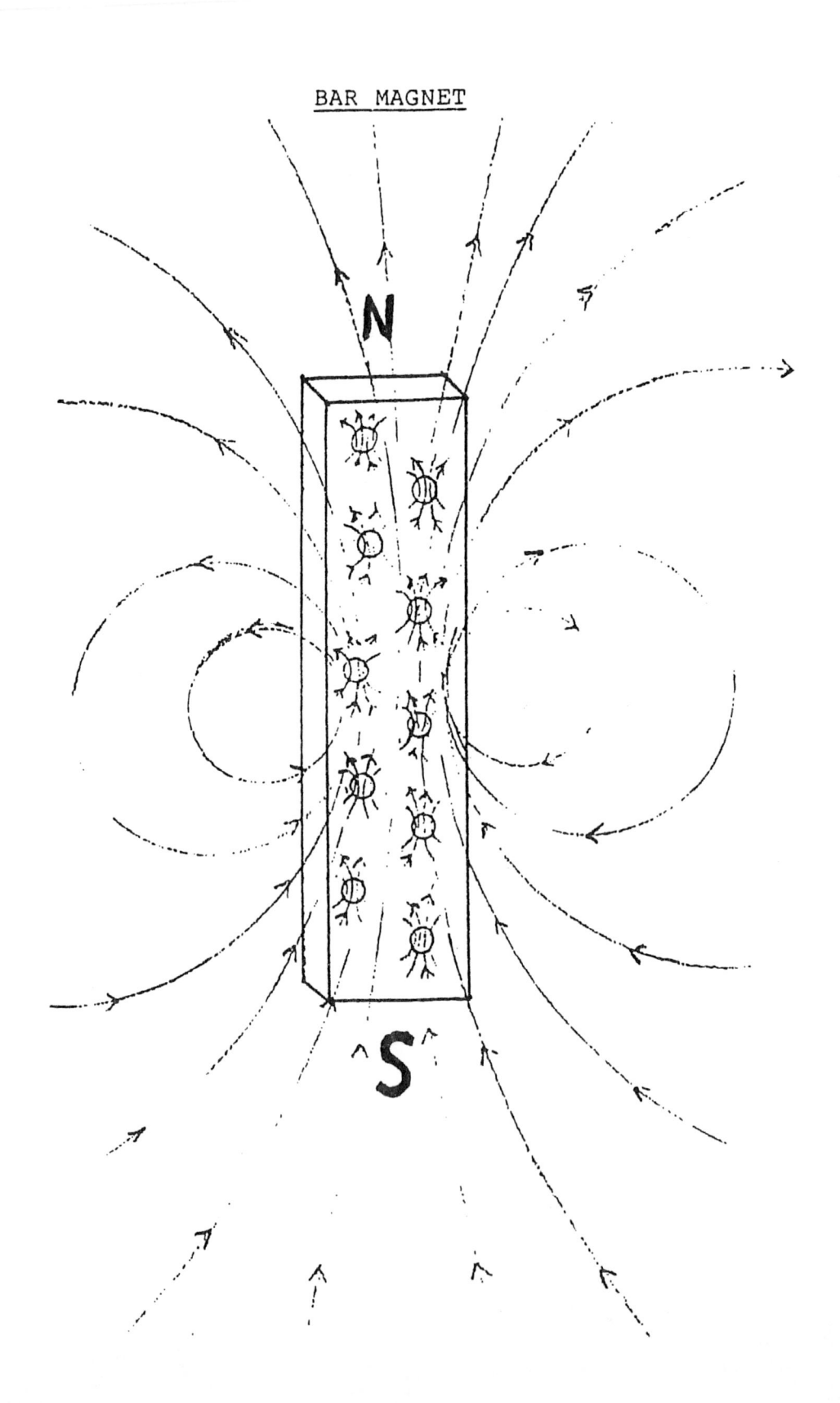

(Continued from page 74)

The atoms in the magnet, many of which are aligned in the same direction, are the "pumps" which produce this "apparent" motion of the "physical substance", making it look to us like it is moving in a complete circuit (see pages 75, 76, and 78). This is why it would be difficult to isolate a magnetic north pole from its corresponding magnetic south pole.

The "apparent" motion of the "physical substance" can be considered to form a closed circuit. If you break the circuit at any point, the magnetic field ceases to exist. These appearances of the "apparent" motions of the "physical substance" itself are actually produced by "rate of change" distortions in the primordial singularity which are consequently represented by the Permeability value.

These "rate of change" distortions, which are represented by the Permeability value, are what actually causes us to perceive a magnetic field. Also, "rate of change" distortions are always perpendicular to "pressure-related" distortions which are, in turn, represented by the Permittivity value.

Thus, electric charges and magnetic fields are always accelerated perpendicular to each other when they are under the influence of velocity-related space-time distortions.

(Continued on page 79)

MAGNETIC FIELDS

INDUCED CURRENT

magnetic field

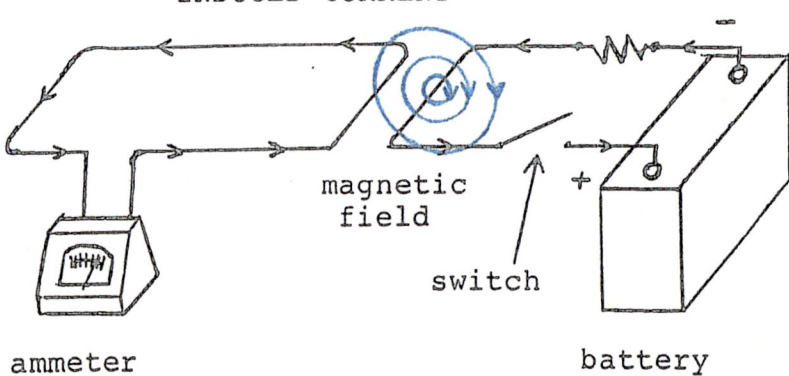

switch

ammeter

battery

The current flow which is shown here, represents the <u>actual</u> movement of the electrons through the wire. The <u>apparent</u> current flow would be in the opposite direction.

An electron represents an area of decreased physical "pressure" in four-dimensional space-time (as it exists at the point of "basic reality"). It is held in that condition because of the effects of relativity on its rapidly oscillating negative half-waveform (explained in chapter 4).

When the switch is closed, the electrons in the wire connected to the battery are placed in motion (accelerated) which creates velocity-related space-time distortion within the electrons along their axis of movement. This causes the four-dimensional space-time itself (which has a "real" existence only at the point of "basic reality") in the vicinity of the electrons to undergo acceleration at a right angle to the direction in which the electrons are moving.

The resulting acceleration of the four-dimensional space-time around the wire (magnetic field), in turn, accelerates the electrons in the nearby wire at a right angle to the magnetic field, but in the opposite direction than the electrons which are moving in the wire connected to the battery.

(Continued from page 77)

This is how electromagnetic wave/particles which have an electric charge and an accompanying magnetic field, such as those within atoms, can be used to create both an electric charge and/or a magnetic field.

Consequently, this is how electric charges and magnetic fields can be used to influence charged electromagnetic wave/particles.

Remember though, that the actual physical properties of the "physical substance" have a real existence only at the point of "basic reality".

Whenever you set an electric charge in motion, or accelerate an electric charge, you also create a magnetic field as the electric charge, which is generally a spherical region of pressure-related space-time distortion, begins to experience velocity-related space-time distortion along its axis of travel (relative motion).

This creates two different types of space-time distortion within the same region of space. The result is that the physical substance itself experiences "rate of change" distortions along an axis which is determined by the orientation of the two different types of space-time distortion.

Since the two different types of space-time distortion always have a perpendicular orientation with respect to the "time line", then the actual "physical substance" always experiences a "rate of change" distortion which is perpendicular to the velocity-related distortion that represents the relative motion of an electric charge.

The "rate of change" distortion within the "physical substance", which is perpendicular to the relative motion (velocity-related space-time distortion) of the electric charge, as in an electromagnet, now has the ability to accelerate an electric charge perpendicular to the "rate of change" distortion in the "physical substance" (an electron within a length of wire for instance).

When the magnetic field is accelerating and expanding, however, the magnetic lines of flux, which represent the "rate of change" distortion in the "physical substance", cut across a nearby length of wire at a right angle (see page 78), which accelerates the electrons within the length of wire perpendicular to the direction in which the magnetic lines of flux are cutting across the wire (more velocity-related space-time distortion).

When considering the relative motion of the electrons in a length of wire, and also the "rate of change" distortions of the "physical substance" which produce our perceptions of magnetic fields, both the electrons and the "rate of change" distortions in the physical substance itself, produce the same type of changes in the Permittivity and Permeability values that a spaceship would undergo when considering its relative motion with respect to the Earth.

(Continued on page 82)

ELECTRIC AND MAGNETIC FIELD RELATIONSHIPS

*Note the repetition and similarities in all these values

Speed of light $c = \dfrac{1}{\sqrt{\varepsilon_0 \mu_0}}$ = 299,792,458 meters/second

1/the speed of light $\dfrac{1}{c} = \sqrt{\varepsilon_0 \mu_0}$ = 3.335,640,951,981,52E-9 m/sec

Speed of light squared $c^2 = \dfrac{1}{\varepsilon_0 \mu_0}$ = 898,755,178,736,817,64 m/sec

1/the speed of light squared $\dfrac{1}{c^2} = \varepsilon_0 \mu_0$ = 1.112,650,056,053,62E-17 m/sec

*ESU - Electrostatic System of Units

ESU of Potential = 299,792,458/1.000,000 = 299.792,458 Volts
ESU of Resistance = 299,792,458^2/100,000 = 898,755,178,736.817,64 Ohms
ESU of Current = 1/299,792,458/10 = 3.335,640,951,981,52E-10 Amperes
ESU of Conductance = (1/299,792,458^2) X 100,000 = 1.112,650,056,053,62E-12 Siemens
ESU of Capacitance = 1/299,792,458^2 X 100,000 = 1.112,650,056,053,62E-12 Farads
ESU of Inductance = 299,792,458^2/100,000 = 898,755,178,736.817,64 Henrys
ESU of Quantity = 1/299,792,458/10 = 3.335,640,951,981,52E-10 Coulombs

---
Volt = 1/(299,792,458/1,000,000) = 3.335,640,951,981,52E-03 ESU of Potential
Ohm = 1/(299,792,458^2/100,000) = 1.112,650,056,053,62E-12 ESU of Resistance
Amperes = 1/(1/299,792,458)/10 = 2,997,924,580 ESU of Current
Siemens = 1(1/299,792,458^2 X 100,000 = 898,755,178,736.817,64 ESU of Conductance
Farads = 1/(1/299,792,458^2 X 100,000) = 898,755,178,736.817,64 ESU of Capacitance
Henrys = 1/(299,792,458^2/100,000) = 1.112,650,056,053,62E-12 ESU of Inductance
Coulombs = 1/(1/299,792,458)/10 = 2,997,924,580 ESU of Quantity

---
Electric Potential = Watts/Amperes
Electric Resistance = Volt/Amperes
Electric Current (magnetomotive force) = Amperes
Electric Conductance = Ampere/Volts
Electric Capacitance = Ampere-second/Volts
Electric Inductance = Volt-Second/Amperes
Electric Quantity = Ampere-Seconds

(Continued from page 80)

ε_0 the Permittivity of free space: 0.000,000,000,008,854,187,817,6 Farads per meter

μ_0 the Permeability of free space: 0.000,000,125,663,706,144 Henry's per meter^{-1} (The $^{-1}$ exponent means that there is an inverse relationship between ε_0 and μ_0)

In other words, equal changes in the Permittivity and Permeability values (velocity-related space-time distortion) which produce our perceptions of acceleration and relative motion, $[(T - T')$ and $(L - L')]$ are the same for everything in this universe; material objects, waves in four-dimensional space-time, and even distortions within the "physical substance" itself.

Consequently, an equal difference in the Permittivity and Permeability values $[(T - T')$ and $(L - L')]$ in the atoms of which two different reference frames consist, will cause both of those reference frames to experience acceleration in relation to each other, regardless of what particular situation you may have: a gravitational field, an electric charge (perpendicular), magnetic fields (perpendicular), or two objects with relative motion. This is how all of the forces in our universe are unified within the primordial singularity which still exists within the realm of "basic reality.

As we were able to see in this chapter, the characteristics of four-dimensional space-time such as: electric, magnetic, and gravitational fields, are an essential part of what we see going on around us. The electric charge is a change or alteration in the permittivity constant by itself. A magnetic field is a change or alteration in the permeability constant by itself. And a gravitational field (also relative motion) is a reduction in the effects of both the permittivity and permeability constants together within any given region of space.

Now, in the next chapter, we will attempt to develop a better understanding of electromagnetic waves in four-dimensional space-time and our universe in general.

CHAPTER 4

ELECTROMAGNETIC WAVES IN FOUR-DIMENSIONAL SPACE-TIME

Now that you have a better understanding of the relationship between gravitational, electric, and magnetic fields in four-dimensional space-time, let's take a look at electromagnetic waves in four-dimensional space-time.

In the last chapter, we came to the conclusion that our perceptions of an electric field are related to physical pressure differences in in the "physical substance", and that our perceptions of magnetism are caused by "change", or, "rate of change" distortions in the "physical substance" itself.

Any time you create a disturbance in a physical medium such as water, air, metal, solid rock, or any type of physical substance, wave motion will result. These waves will travel (propagate) through the medium (substance) at a speed, or rate, which is determined by several factors such as: pressure, tension, density, viscosity, elasticity, or hardness.

There are two main characteristics of waves as they propagate through virtually every type of medium. The first of which are the pressure distortions (compressions and rarefactions) of the medium, and the second is the changes, or movements, of the medium itself through which the waves are propagating. Consequently, these principles also apply to the underlying physical process which is producing our perceptions of electric and magnetic fields in addition to electromagnetic waves.

The electric charge represents the pressure variations in the "physical medium", and the magnetic field represents the "apparent" motion, or "change" in the "fixed state", of the physical substance itself within the primordial singularity. Due to the fact that these two types of distortions in the physical medium are connected and simultaneous with each other, in addition to being perpendicular to each other as we saw in the previous chapter, this is why waves in four-dimensional space-time are called "electro-magnetic" waves (see page 95).

Electromagnetic waves also come in two basic types: The first type is a **transverse wave**, which can be compared to the up and down motion of a wave on a tight string. This type of wave motion is always perpendicular to the direction in which the wave is travelling, or propagating. The second type is known as a **longitudinal wave**, or compression wave in which the compressions and de-compressions, of the medium in which the wave exists, lie along the same axis as the propagation of the wave.

Even though electromagnetic waves can have both, a transverse or a longitudinal polarization, the transverse wave polarization of electromagnetic waves are what we find most interesting because laboratory experiments have shown that transverse waves are only able to propagate in a solid substance.

Laboratory experiments performed many years ago have shown that longitudinal waves can propagate through both, a solid or a liquid, but that transverse waves can only propagate through a solid because there has to be a "restoring", or "holding", force within the medium to enable them to propagate. In other words, the conducting medium has to be held stationary, or in a "fixed state", in order to enable the transverse waves to propagate. As a result, this evidence was originally used to determine that the core of the Earth is molten.

During an earthquake, both longitudinal ("P" waves) and transverse ("S" waves) are produced and can be detected at the surface because both types of waves can travel through the Earth's solid crust. However, due to the fact that there is no restoring (holding) force in the liquid (molten) core, the transverse waves are unable to travel through the Earth's core. The longitudinal waves, on the other hand, are able to travel right on through the molten core and can be detected on the opposite side of the Earth.

This evidence makes a convincing argument that the "physical substance" in the primordial singularity is being held in a "fixed state" by the laws of physics of the void in which it exists (realm of "basic reality"). As a result, our perceptions about electromagnetic waves are being produced by a combination of "pressure" and "change" types of distortions in this "physical substance" as it is being held "fixed" or stationary (see page 200 in the Appendix).

The fact that Albert Einstein presented this evidence in one (perhaps even more) of his lectures makes this whole situation truly remarkable because the "Theory of Relativity" is what was *supposed* to have made the previous "ether theory" obsolete. Surprisingly, instead of disproving it, the Theory of Relativity provides us with some of the most convincing evidence which *supports* the "ether theory" in a somewhat modified form.

This evidence includes the mathematical description of the physical expansion process by the "time" and "length" equations (see page 15), and now we are seeing even more convincing scientific evidence that the "physical substance" is being held in a "fixed state" by the nature of the void in which it exists.

This set of circumstances can also explain why we have "constants" in physics.

(Continued on page 90)

HARMONIC WAVE MOTION

$$v = v_m \sqrt{1 - \frac{x^2}{A^2}}$$

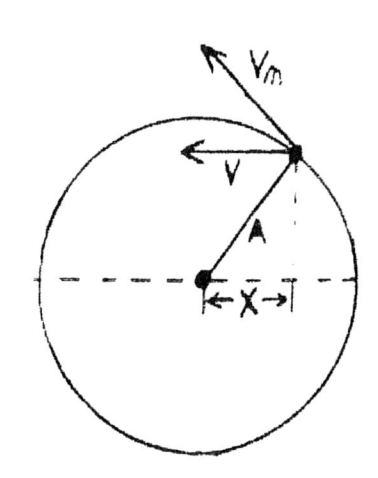

This drawing can be used to provide a two-dimensional representation of the process by which electromagnetic waves (kinetic and potential energy) are related to uniform circular motion.

If you can imagine that the point at which lines Vm, V, A, and the vertical dotted line intersect was slowly rotated around the circle at 1 revolution per second, you would see the vertical dotted line moving back and forth along the horizontal dotted line in a manner which represents wave oscillation at a rate of 1 full cycle per second. The point at which Vm, V, and A, intersect can then be speeded up to represent any frequency (rate) of oscillation. "X" represents the actual displacement (compression), both positive or negative, at any point in time, of the medium or substance that these waves (vibrations) are in.

The next drawing can be used to represent harmonic oscillations in any type of medium or substance including the Permittivity (electric charge) and Permeability of free space (magnetic field) of which electromagnetic waves consist. You should always remember, however, that these drawings are a two-dimensional representation of a three-dimensional concept.

(Harmonic Wave Motion)

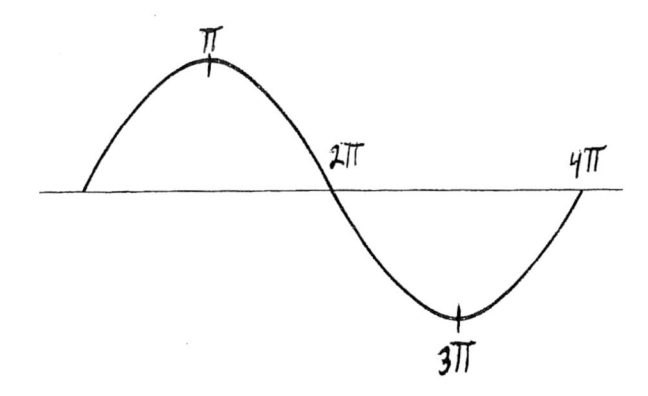

KE = ½ mv²   PE = ½ KX²   E = ½ KA²

WAVE PROPAGATION RATES

For longitudinal waves in an ordinary substance or liquid:

$$\text{velocity} = \sqrt{\frac{\text{elastic force factor, or, bulk modulus}}{\text{density, or, inertia factor}}}$$

For transverse waves in an ordinary solid substance:

$$\text{velocity} = \sqrt{\frac{\text{pressure or tension}}{\text{mass per unit length}}}$$

For electromagnetic waves:

$$\text{velocity} = \frac{1}{\sqrt{\text{Permittivity X Permeability of free space}}}$$

There are other relationships which are also common to these different types of waves which include:

$$\text{velocity} = \frac{\text{wavelength}}{\text{time}}$$

$$\text{velocity} = \text{wavelength X frequency}$$

(Continued from page 87)

The Lorentz transformation equations, which Albert Einstein used in his Special Theory of Relativity, are used throughout this theory because the curve (on a graph) that they plot describes the underlying physical process (expansion) of the "physical substance". The expansion of the "physical substance" is the underlying process which produces our four-dimensional space-time continuum (the universe that we perceive) (see pages 15 and 71).

The total energy which is transmitted, or represented, by an electromagnetic wave can be calculated by multiplying the relative strengths of the electric charge and magnetic field which are then multiplied by 2π to give you the total potential energy which is represented by the amplitude/angular momentum, and then by multiplying that times the frequency of the wave.

The amplitude/angular momentum represents the relative strength of the electric charge times the relative strength of the magnetic field which is then multiplied times 2π because electromagnetic waves undergo a type of rolling, or twisting, motion as they propagate through space.

This rolling, or twisting motion, of the electric and magnetic fields is called angular momentum (spin) and it represents a significant portion of the potential energy which is carried by the electromagnetic wave (more on this in chapter 5).

The energy contained in the amplitude/angular momentum is then multiplied times the number of individual wavelets passing a certain point per second (frequency) which gives you the total amount of energy which is carried by a photon.

However, we should remember, when considering these relationships, that we live in a universe which, as we see it, behaves according to Einstein's Theory of Relativity -- velocity-related space-time distortion -- which is always perpendicular to the actual physical pressure-related and "change"-related distortions within the primordial singularity.

Due to the inverse relationship (see page 89) regarding the propagation of electromagnetic waves as compared to waves in other substances, electromagnetic waves do propagate in a rather strange and unusual manner. There are various quirks and oddities regarding the transmission and reception of electromagnetic waves which require us to regard them in a somewhat non-classical fashion.

With these oddities being taken into account, if you could actually see it, an electromagnetic wave would "appear" to us like an expanding sphere moving outward in all directions from its "point of origin" at the speed of light, 299,792,458 meters per second (see page 20). A three-dimensional electromagnetic wave is quite similar to the type of wave disturbance which would be created by striking a very large, solid metal object with a hammer. These types of shock waves are somewhat comparable to basic electromagnetic waves in four-dimensional space-time.

Again, the velocity at which a wave propagates in any type of medium or substance is determined by the physical properties of the medium itself such as: elasticity, viscosity, pressure, stress, tension, or hardness.

Since our perceptions of relative motion are produced by "changes" in both the Permittivity and Permeability values, equally, and since an electric field is a pressure-related distortion in the "physical substance" (an increase or decrease in the Permittivity value by itself), and a magnetic field is produced by a "change of position" type of disturbance in the "physical substance" resulting in an increase or decrease in the Permeability value by itself, then our observations of electromagnetic waves will always appear to us to be somewhat different than the "apparent" wave motion which is represented by the "physical substance" within the primordial singularity.

On the surface of an electromagnetic wave, we have the initial high pressure area of the wave in which the actual physical substance is being compressed or squeezed together (see page 95).

In order for the leading edge of the wave to build up pressure, the physical substance "changes" (magnetic north pole) to a more highly compressed state which coincides with the compression of the substance to a higher physical "pressure" (positive electric charge).

(Continued on page 96)

ELECTROMAGNETIC RADIATION SPECTRUM

(HIGHER FREQUENCIES)

CHARACTERISTICS	FREQUENCY	WAVELEGNTH	ENERGY	MASS

As the frequency continues upward above this line, the space-time distortion surrounding the wave/particle (gravitational field) as a result of its extremely rapid wave motion, increases to such a degree that other wave/particles are drawn inside or "captured". This extremely energetic wave/particle becomes what is known as the "singularity" at the center of a black hole.

Psi wave/particle	8.9×10^{23} Hz	4×10^{-14} cm	3685 MeV / 5.89×10^{-3} erg	6.45×10^{-16} g
		10^{-13} cm		
neutron	2.27×10^{23} Hz	2×10^{-13} cm	939.6 MeV	1.675×10^{-24} g
proton			938.3 MeV	1.673×10^{-24} g
	10^{23} Hz			
		10^{-12} cm		
+/- pion	3.26×10^{22} Hz	1.75×10^{-12} cm	139.6 MeV	2.4×10^{-25} g
+/- muon	2.25×10^{22} Hz	2.5×10^{-12} cm	105.7 MeV	1.89×10^{-25} g
	10^{22} Hz			
		10^{-11} cm		
	10^{21} Hz			
		10^{-10} cm		
gamma rays	1.2×10^{20} Hz	3.9×10^{-10} cm	5.11 MeV / 8.2×10^{-7} erg	9.11×10^{-28} g
(electron/positron)	10^{20} Hz			

- -

Above this frequency, the electro-magnetic wave/particles begin to experience the effects of relativity as a result of their high rate of oscillation (wave motion), in addition to their rate of propagation, thus, $E = M \times C^2$ with C^2 representing the propagation rate times the rate of oscillation.

	10^{19} Hz			
		10^{-8} cm		
X - rays	3×10^{18} Hz	1.75×10^{-8} cm	9.88×10^{-9} erg	1.1×10^{-29} g
	10^{18} Hz			
		10^{-7} cm		
	10^{17} Hz			
		10^{-6} cm		
ultra-violet radiation	10^{16} Hz	$.5 \times 10^{-5}$ cm	6.6×10^{-11} erg	7.3×10^{-32} g
		10^{-5} cm		

↓ (TO LOWER FREQUENCIES) ↓

ELECTROMAGNETIC RADIATION SPECTRUM

(LOWER FREQUENCIES)

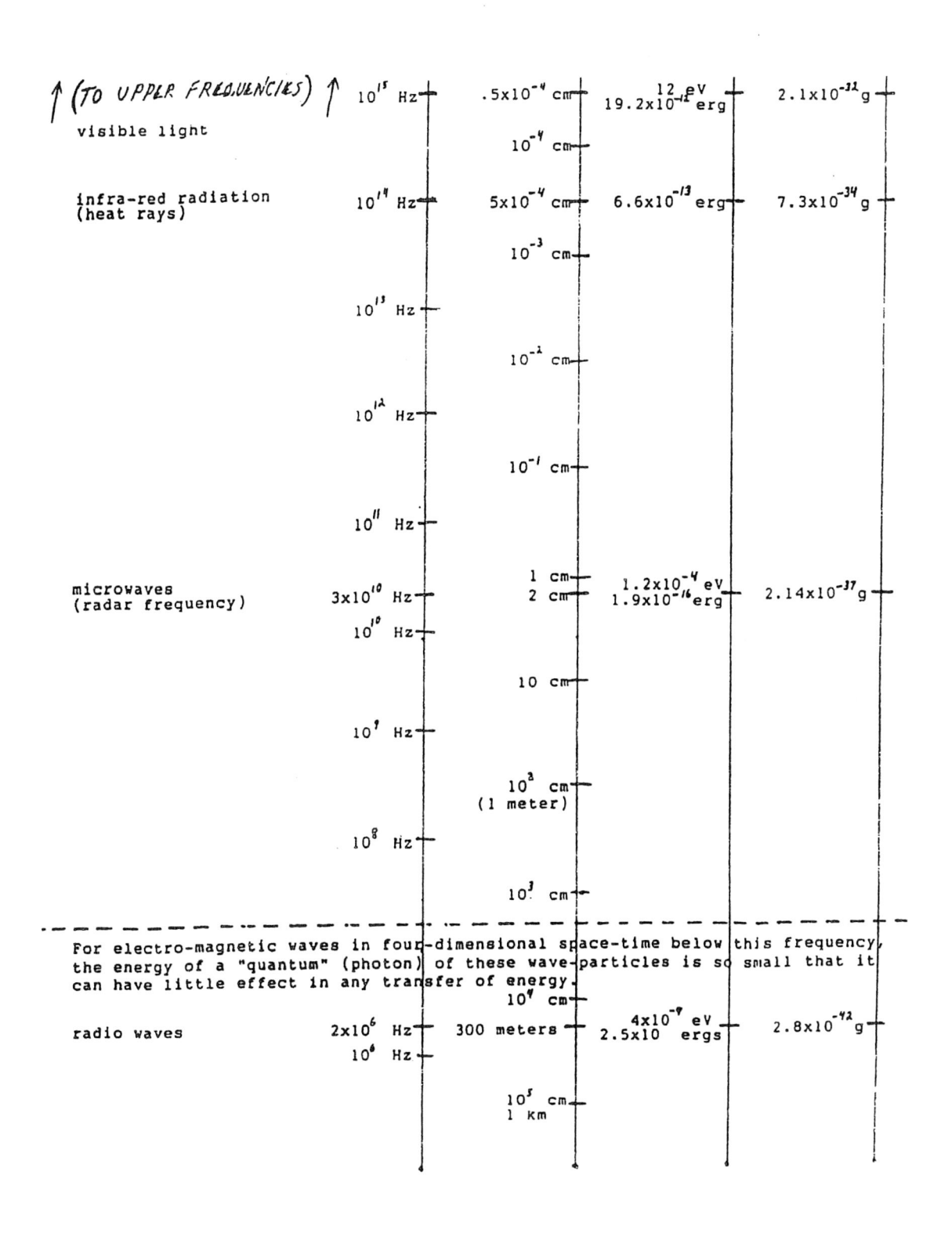

↑ (TO UPPER FREQUENCIES) ↑ 10^{15} Hz $.5 \times 10^{-4}$ cm 19.2×10^{-12} erg, 12 eV 2.1×10^{-32} g

visible light

          10^{-4} cm

infra-red radiation 10^{14} Hz 5×10^{-4} cm 6.6×10^{-13} erg 7.3×10^{-34} g
(heat rays)

          10^{-3} cm

     10^{13} Hz

          10^{-2} cm

     10^{12} Hz

          10^{-1} cm

     10^{11} Hz

microwaves   1 cm 1.2×10^{-4} eV 2.14×10^{-37} g
(radar frequency) 3×10^{10} Hz   2 cm 1.9×10^{-16} erg

     10^{10} Hz

          10 cm

     10^{9} Hz

          10^{2} cm
          (1 meter)

     10^{8} Hz

          10^{3} cm

- -

For electro-magnetic waves in four-dimensional space-time below this frequency, the energy of a "quantum" (photon) of these wave-particles is so small that it can have little effect in any transfer of energy.

          10^{4} cm

radio waves 2×10^{6} Hz   300 meters 4×10^{-9} eV 2.8×10^{-42} g
     10^{6} Hz 2.5×10 ergs

          10^{5} cm
          1 km

PHOTON

(two-dimensional view)

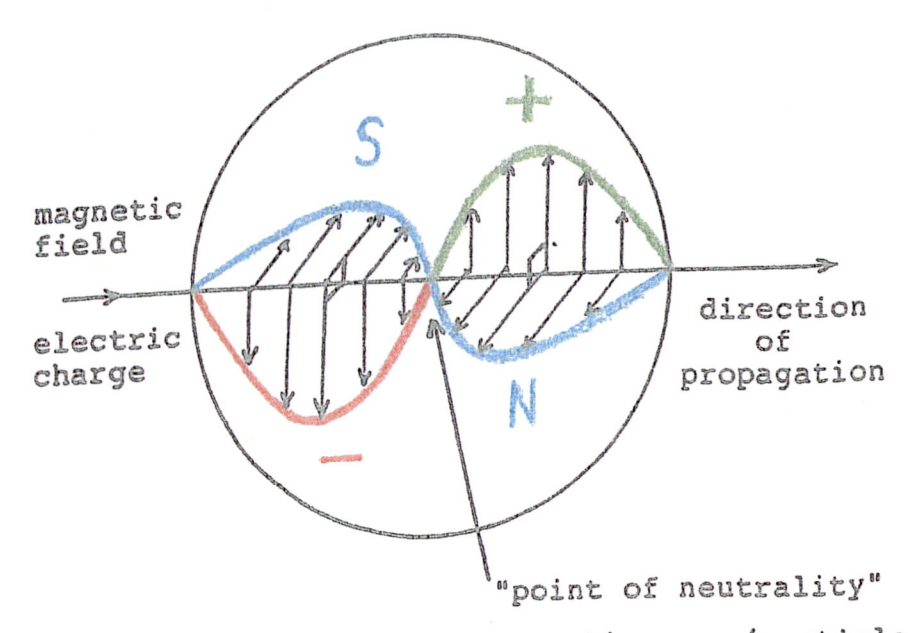

[Diagram: a circle containing a sinusoidal wave along a horizontal axis labeled "direction of propagation". The axis entering from the left is labeled "magnetic field" and "electric charge". The upper half of the wave is marked with "S" and "+"; the lower half with "N" and "−". A line points to the wave crossing the axis labeled "point of neutrality".]

    A photon is an electro-magnetic wave/particle in four-dimensional space-time which has a dual existence... both wave and particle (as we see it). Its wave characteristics (wave function) travel outward in all directions at the speed of light (186,282 miles per second) like an expanding sphere. Its particle characteristics represent the electro-magnetic wave as it exists at the point of "basic reality" (primordial fireball singularity) on the Time Chart.
    The point of "basic reality" (on the Time Chart) represents our universe as it exists in an otherwise empty and endless black void where time stands still (realm of "basic reality"). Our universe <u>still</u> exists there as a single point (primordial fireball singularity) as if the "big bang" has not occurred yet.
    As we have seen earlier, the physical (electro-magnetic) properties of four-dimensional space-time exist only at the point of "basic reality", and the "pressure" of four-dimensional space-time inside the primordial fireball singularity was very, very high - so high, that electro-magnetic waves (photons) travel through it at a very high rate of speed (186,000 miles per second).

(Continued from page 92)

The physical substance then "changes" (magnetic south pole) to a more rarified state which coincides with the area of low "pressure" (negative electric charge) which represents the negative half of the wave. Also, the electric charge and the magnetic field, within an electromagnetic wave, are always perpendicular to each other because they represent the two different types of distortions in the same physical medium within the primordial singularity.

Electromagnetic waves will still be three-dimensional, even though their shape may vary and they may be directed more in one direction than another because of the possibility of slightly different orientations regarding the positive electric charge (high pressure) and the negative electric charge (low pressure) portions of the wave.

The frequency of an electromagnetic wave is the number of full-cycle oscillations passing a certain point within a given time span. Frequency is usually measured in cycles per second, or Hertz, named after Heinrich Hertz (1857 – 1894) who was the first physicist to successfully transmit and receive electromagnetic waves back in 1887.

The greater the disturbance in the physical substance which creates an electromagnetic wave, the stronger the electric charge ("pressure" difference) and the corresponding magnetic field ("change" in the physical substance) will be. The energy contained in the strength of the electric charge is usually equal to the energy contained in the strength of the magnetic field, and by multiplying them together, along with 2π, you can determine the total amplitude/angular momentum of the wave (see page 119).

Due to the fact that It requires more energy to produce a faster "rate of change" in the physical substance, between high pressure (+) and low pressure (-) than it does for an electromagnetic wave which has a slower "rate of change". It stands to reason then, that it requires more energy to produce a more rapidly oscillating (higher frequency) electromagnetic wave than it does for a slower-oscillating wave.

The frequencies of electromagnetic waves vary widely, from very low frequencies to extremely high frequencies on the order of 10^{23} Hz. The wavelength, which represents the distance between wave crests, is inversely proportional to the frequency, which means that as one value gets larger, the other value gets smaller, and vice versa, which is also representative of the space-time inverse relationship (see page 15).

Since we know that electromagnetic waves propagate through our space-time continuum at the rate of 299,792,458 meters per second, an electromagnetic wave with a frequency of one cycle per second, 1 Hz, would have a wavelength of 299,792,458 meters. An electromagnetic wave with a frequency of ten cycles per second, or 10 Hz, would have a wavelength of 299,792,458 meters divided by 10 which equals 29,979,245.8 meters.

The Permittivity and Permeability values of free space produce our perceptions about both the "pressure-related" and "velocity-related" distortions associated with electromagnetic waves.

ε_0 the Permittivity of free space: 0.000,000,000,008,854,187,817,6 Farads per meter

μ_0 the Permeability of free space: (0.000,000,125,663,706,144 Henry's per meter^{-1} (The $^{-1}$ exponent means that there is an inverse relationship between ε_0 and μ_0)

The expansion of our universe, our concepts of time and space, cause us to perceive an electromagnetic wave which appears to us as though it is propagating (expanding and spreading out) through our four-dimensional space-time continuum at a rate of 299,792,458 meters per second.

As we have seen previously, the rate at which a wave propagates through a substance or medium is determined by its pressure, tension, or hardness. Can you just imagine how much pressure or tension the "physical substance" would have to be under for wave disturbances in it to appear to us as though they propagate at such a high rate of speed?

This situation would lead us to conclude that the primordial singularity is still under a tremendous amount of "pressure" just as it would have been the instant before it began expanding and spreading out to form the universe that we perceive, according to the time and length equations in the Special Theory of Relativity. This would also lead us to conclude that the primordial singularity is being held in this highly pressurized state by the nature of the realm, or void, in which it exists which is referred to as the realm of "basic reality" in this theory.

In Chapter 2, we discussed relativity, and how time slows down and length contracts for something travelling at a high relative velocity such as 90% of the speed of light for example. But what about something travelling "at" the speed of light (299,792,458 meters per second)?

Since the relativity theory states that time slows down and length contracts for anything travelling at a relative velocity which approaches the speed of light, then for something travelling "at" the speed of light, time should completely stop and length would contract down into (nearly) a single point in space.

Since an electromagnetic wave, if it could be seen, would look to us like a sphere expanding outward in all directions at the speed of light, then the length contraction would occur in all directions so that the whole sphere would be contracted down to nearly a single point in space. An electromagnetic wave (or series of waves) with a frequency below 10^{20} Hz, which is contracted into a point/particle in this manner, has become known to physicists as a photon (see page 95).

This is what gives electromagnetic wave/particles, photons, neutrons, electrons, protons, etc., a dual existence, as we see it, both wave and particle.

For many years physicists tried to determine, conclusively, whether electromagnetic waves (light) consisted of waves or particles. Experimental evidence seemed to favor first one side and then the other. In 1801, an English physicist named Thomas Young (1773 – 1829) devised an ingenious experiment which, at first, appeared to settle the argument, decisively, in favor of electromagnetic waves. He conducted his experiment by directing a beam of sunlight onto a screen with two parallel slits cut into it a few millimeters apart.

When the beam of sunlight entered the two slits, a wave interference pattern appeared on another screen which was placed behind the slits, thus, proving that the electromagnetic waves were acting like waves and not like particles. Soon afterward, however, it was discovered that when one of the slits was closed off, the electromagnetic waves (light) began acting as if they were particles (photons) again.

This same experiment was conducted later on with high-frequency electromagnetic waves (above 10^{20} Hz) which include electrons, protons and neutrons, with the same result, thus, proving that they exist as both electromagnetic waves and particles, simultaneously.

The controversy surrounding the wave-particle duality of electromagnetic waves continued until the much-more comprehensive theories of Quantum Physics and Quantum Mechanics, developed in the early 1900's, finally settled the argument in favor of a wave/particle duality for electromagnetic waves at all frequencies.

The electromagnetic wave characteristics of a photon, as it exists in our space-time reference frame, are simply an electromagnetic wave, or a whole series of electromagnetic waves --- more on this later, in four-dimensional space-time expanding outward from its starting point (point of origin) in all directions at the speed of light like an expanding sphere (see page 20).

In this theory, this concept is referred to as its "sphere of influence". Since the "effects" of four-dimensional space-time (represented by the Permittivity and Permeability values) are what give us the impression that our universe is spatially extended to its present size and that time passes as we know it, then, with the effects of four-dimensional space-time being totally distorted, or no longer having any "effect", our perceptions of space and time would be meaningless.

A photon could be emitted from an electron on the Sun and propagate a billion kilometers through our solar system and then be absorbed by another electron on a distant object. From the photon's particle point of view, moving from one electron to another would be like stepping through a door way from one room into the next.

No time would have elapsed, and the photon would not realize that any distance had been travelled. Since electromagnetic waves propagate at the speed of light, we know for sure that the real physical characteristics of electric and magnetic fields exist only within the primordial singularity.

Since the "effects" of our four-dimensional space-time energy field are totally distorted with respect to the photon, the electromagnetic wave characteristics of the photon (as we perceive them) appear to us as if they propagate at a very high rate of speed, as if the photon were still in the primordial singularity where the "pressure" of the physical substance is still extremely high, even though the "pressure" of the physical substance in our time frame (as it exists in other "Dimensions"), is much, much lower (see pages 42 and 136). The way that this reduced pressure reveals itself in our time frame is through the relationship between acceleration, relative motion, and gravity (see pages 15 and 71).

In the realm of "basic reality", the concepts of space and time are meaningless, so this causes the singularity which represents our universe, to exist there as if it is held "fixed" in its original state (primordial singularity) as seen by an observer in the realm of "basic reality".

To an observer within our universe, the effects of the four-dimensional space-time energy field on the outside of our universe do not exist (are totally distorted) with respect to the space within our universe (where he or she is). This causes the observer within our universe to perceive that our universe is expanding outward in all directions at an extremely high velocity.

Since electromagnetic waves have a real existence only within the primordial singularity, they are also held "fixed" in their original state, according to the observer in the realm of "basic reality". But according to the observer in our universe, electromagnetic waves appear as if they are expanding outward in all directions at the speed of light.

This difference in opinion between the two observers is why the speed of light, which represents total space-time distortion, remains constant, while the pressure of the "physical substance" is gradually decreasing as our universe continues to expand and spread out.

This effect is also what causes the dual existences of photons, neutrons, protons and electrons --- both wave and particle, simultaneously. Since electromagnetic wave/particles exist in both states simultaneously, whether we observe them as waves or particles can be influenced merely by interactions with measuring devices, and even our knowledge about them (quantum theory) more on this in chapter 5.

The characteristics of electromagnetic wave/particles differ widely at various frequencies (see pages 93 and 94). Low-frequency radio waves have such low-energy photons that they are barely detectable and are not able to participate in any type of energy exchange between other electromagnetic wave/particles.

As you go up the frequency scale, however, you can see that electro-magnetic wave/particles have much different characteristics at the higher frequency (energy) levels. But they are all electromagnetic wave/particles existing as both wave and particle, simultaneously.

Toward the top of the scale, at frequencies in the neighborhood of 10^{20} Hz, we come to a major threshold. The oscillation rate of the electromagnetic wave/particles, at that frequency, becomes so rapid that they contract into particles due to the fact that the extremely high "rate of change" of the physical substance is reaching relativistic rates. Consequently, these extremely energetic electromagnetic wave/particles can now exist at relative velocities which are less than the speed of light.

The waveforms of which these high-frequency wave/particles consist, have such an extremely high "rate of change" between areas of positive and negative "pressure-related" space-time distortion, that this causes the waveform to approach the primordial singularity on the "Time Chart" (see the graph on page 111).

Electromagnetic wave/particles which are contracted into a wave/particle due to the rate at which they propagate in addition to their high oscillation rate, have mass (inertia or weight). According to this theory, Albert Einstein's famous equation $E = mc^2$ describes the fundamental concept that an electromagnetic wave/particle, which consists of the amount of energy "E", is actually experiencing the maximum effects of relativity twice.

First, is its propagation rate which is "c" (the speed of light), and secondly, is its rate of oscillation (rate of change), which for wave/particles which oscillate above the frequency threshold near 10^{20} Hz, is also "c". In other words, the propagation rate (c) multiplied times the oscillation rate (c) is equal to c^2. Wave/particles which oscillate above 10^{20} Hz are experiencing the maximum effects of relativity (total space-time distortion) twice, and thus, have mass.

(Continued on page 106)

ELECTROMAGNETIC WAVE/PARTICLE DECAY

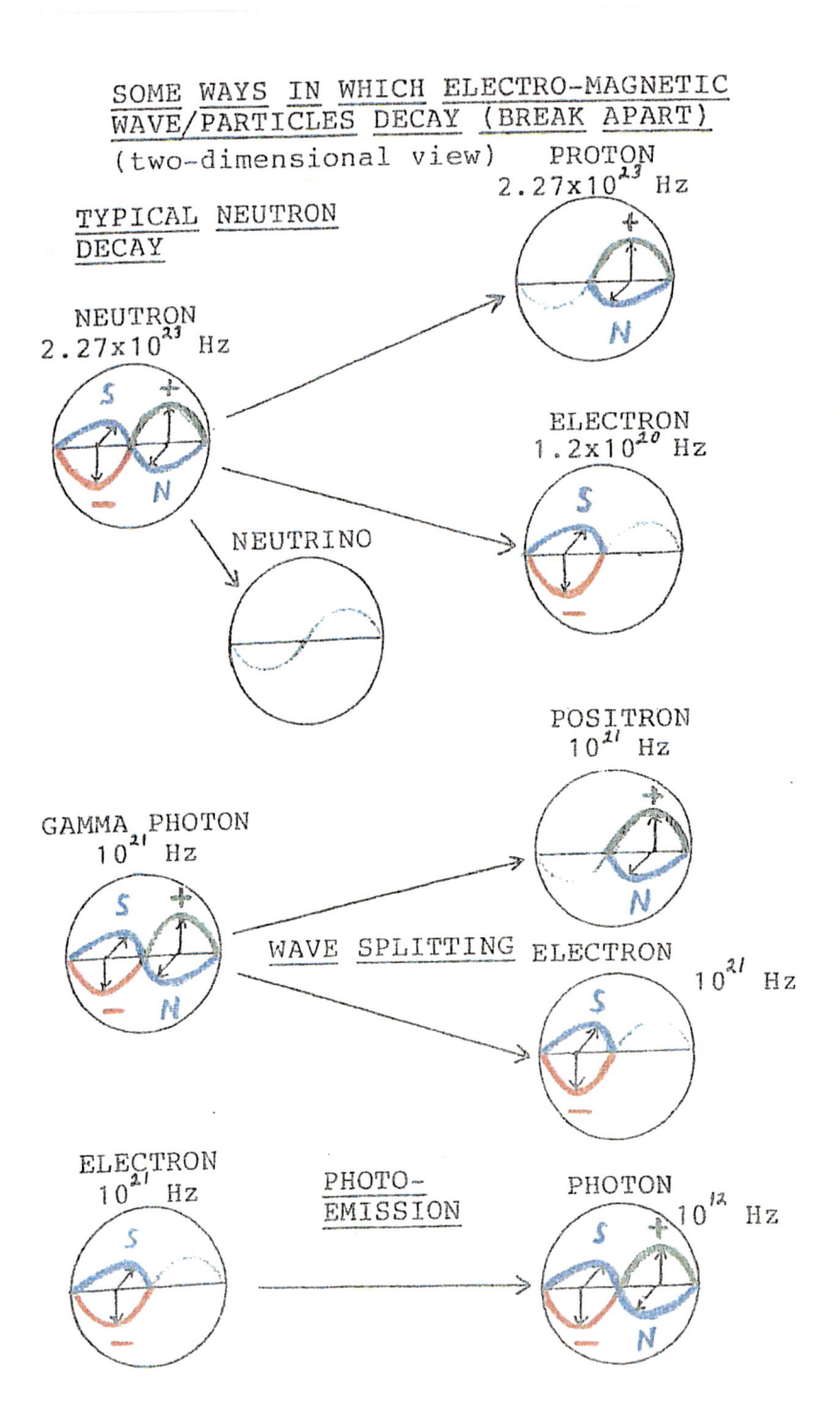

SOME WAYS IN WHICH ELECTRO-MAGNETIC WAVE/PARTICLES DECAY (BREAK APART)

(two-dimensional view)

TYPICAL NEUTRON DECAY

NEUTRON
2.27×10^{23} Hz

→ PROTON 2.27×10^{23} Hz

→ ELECTRON 1.2×10^{20} Hz

→ NEUTRINO

GAMMA PHOTON
10^{21} Hz

WAVE SPLITTING

→ POSITRON 10^{21} Hz

→ ELECTRON 10^{21} Hz

ELECTRON 10^{21} Hz

PHOTO-EMISSION

→ PHOTON 10^{12} Hz

(Continued from page 104)

This direct correlation between space-time distortion and electromagnetic wave energy can be demonstrated by replacing "c" and "c²" with the corresponding Permittivity (ε_0) and Permeability (μ_0) values in our energy equations and also in our time and length equations (see page 30). With greater and greater levels of space-time distortion, the Permittivity and Permeability values in the denominator become smaller and smaller.

The smaller the denominator becomes in the equation, the greater the numerical value which is represented by factoring the numerator and the denominator, and as a result, the more velocity-related space-time distortion (relative motion) the electromagnetic wave has, and consequently, the more energy (mass) it has.

Another interesting property of electromagnetic wave/particles which oscillate above the frequency threshold near 10^{20} Hz, is that they have a tendency to decay or break apart. (see page 105). There are primarily two ways in which a wave/particle can decay (break apart). The first of these is "Frequency Splitting" which is the emission of a photon by a wave/particle such as an electron. The wave/particle creates a disturbance (another wave, or series of waves) in the four-dimensional space-time energy field in order to release some excess energy.

Now, let's take a look at the "Frequency Splitting" which an electron will undergo for an example. The new wave/particle (photon) will have the same electromagnetic wave amplitude/angular momentum as the electron wave/particle which emitted it (Planck's constant) (see pages 119, 120, and 128).

It also has a frequency which is directly proportional to the amount of energy that the electron wave/particle released (see page 105). In other words, the electron creates a photon with the same amplitude/angular momentum that it has, and a frequency which is equal to the reduction in the number of cycles per second (Hz) that the electron wave/particle underwent when it released the photon.

Whenever an electron changes energy states, or the nucleus of an atom releases energy, it creates a disturbance within the four-dimensional space-time energy field, which is represented by the Permittivity and Permeability values, that is quite similar to dropping a stone into a pool of water. Not just one, but a rather large number of individual wavelets are produced -- usually between 100,000 and 1,000,000 for electromagnetic waves.

The sum of all the wave energy which is transferred by all of the individual wavelets, which were produced from the instant that the stone hit the water until the pool became still again, equals the total amount of energy which was released by the stone hitting the water. With electromagnetic waves which are produced by atoms, or the component parts of atoms, the amplitude/angular momentum of the individual wavelets are all the same (Planck's Constant) (see pages 119, 120, and 128). The frequency (f), in oscillations per second of the individual wavelets, are what actually determines how much energy (E) is transferred by the total energy release.

Earlier, we discussed how electromagnetic waves experience the maximum effects of relativity as a result of their propagation rate, and how this causes the electromagnetic waves to be contracted into a point particle. Since all of the electromagnetic wavelets which are produced by each quantum event, are also created at the same point of origin (location) and are all created within a certain time period, then they are all contained within the same wave/particle (photon) which can now be represented by the formula: E = h f

With the frequency (f) being measured in oscillations per second you will need to multiply (f) times the actual length of time (in seconds) that it took for the medium to calm back down again. Since Planck's constant (h) is a known quantity (see page 119), and if you know the total amount of energy (E) which is being transferred by the photon, then you can simply divide (E) by Plank's constant (h) to calculate the total number of individual electromagnetic wavelets which are represented by an individual photon.

The dual existence of the photon, which is produced by the effects of our four-dimensional space-time energy field, causes the photon (particle characteristics) to appear to us as though it exists simultaneously at all points on its wave-front (wave characteristics).

When the outer edge of the wave-front encounters an electron in an atom, for instance, and the photon is absorbed by the electron, the whole electromagnetic wave front collapses instantly (wave function collapse) so that the electron receives all the wave energy that created the photon in the first place, and very little is lost, even if the electromagnetic wave-front has expanded to include a sphere of influence hundreds of millions of kilometers in diameter! This is how energy is transferred from atoms on the Sun, to a destination in an atom on the Earth through millions of kilometers of space.

Another way in which an electromagnetic wave/particle can decay (break apart) is by "wave-splitting" (see drawing on page 105). A neutral electromagnetic wave/particle which contains both the negative and positive wave-halves, such as a gamma photon or a neutron, contains a weak "point of neutrality" between the two wave-halves.

These two half-waveforms can then become separated from one another and can form an electron (negative half-wave) and a positron (positive half-wave) in the case of a Gamma wave/particle, or an electron and a proton in the case of a neutron wave/particle (see page 105). This process is a commonly observed quantum event in hydrogen bubble chamber experiments in particle physics laboratories. This process is the same, regardless of the frequency above the threshold near 10^{20} Hz at which the wave/particles are oscillating (see page 93).

Another observation which can be presented here is that all of the mass in our universe appears to be the product of neutron decay, in which free neutrons decayed to form protons and electrons which then recombined to form hydrogen and helium atoms. No other subatomic wave/particles occur naturally in large enough numbers to have been able to influence the neutrons or the production of atoms and molecules (mass) in any meaningful way. All of the atoms and molecules in our universe are composed of just three basic components: neutrons, protons, and electrons.

The frequency of the wave/particles determines their mass and physical characteristics. This indicates that everything in our universe: light; matter; cosmic rays; mass; etc.; consists entirely of electromagnetic waves, or the component parts of electromagnetic waves at various frequencies.

According to this theory, an electromagnetic wave/particle is a space and time distortion wave. Since we have seen previously that space and time are closely related, that is, you cannot change one without affecting the other, then there must be a space component and a time component in a space-time distortion wave.

The frequency is the time component and the wavelength is the space component of the electromagnetic wave. As the frequency increases toward the upper end of the graph on page 111, there is such an extremely high "rate of change" between areas of positive and negative "pressure-related" space-time distortion, that this also produces significant space-time dilation within the immediate vicinity of the wave/particle.

(Continued on page 112)

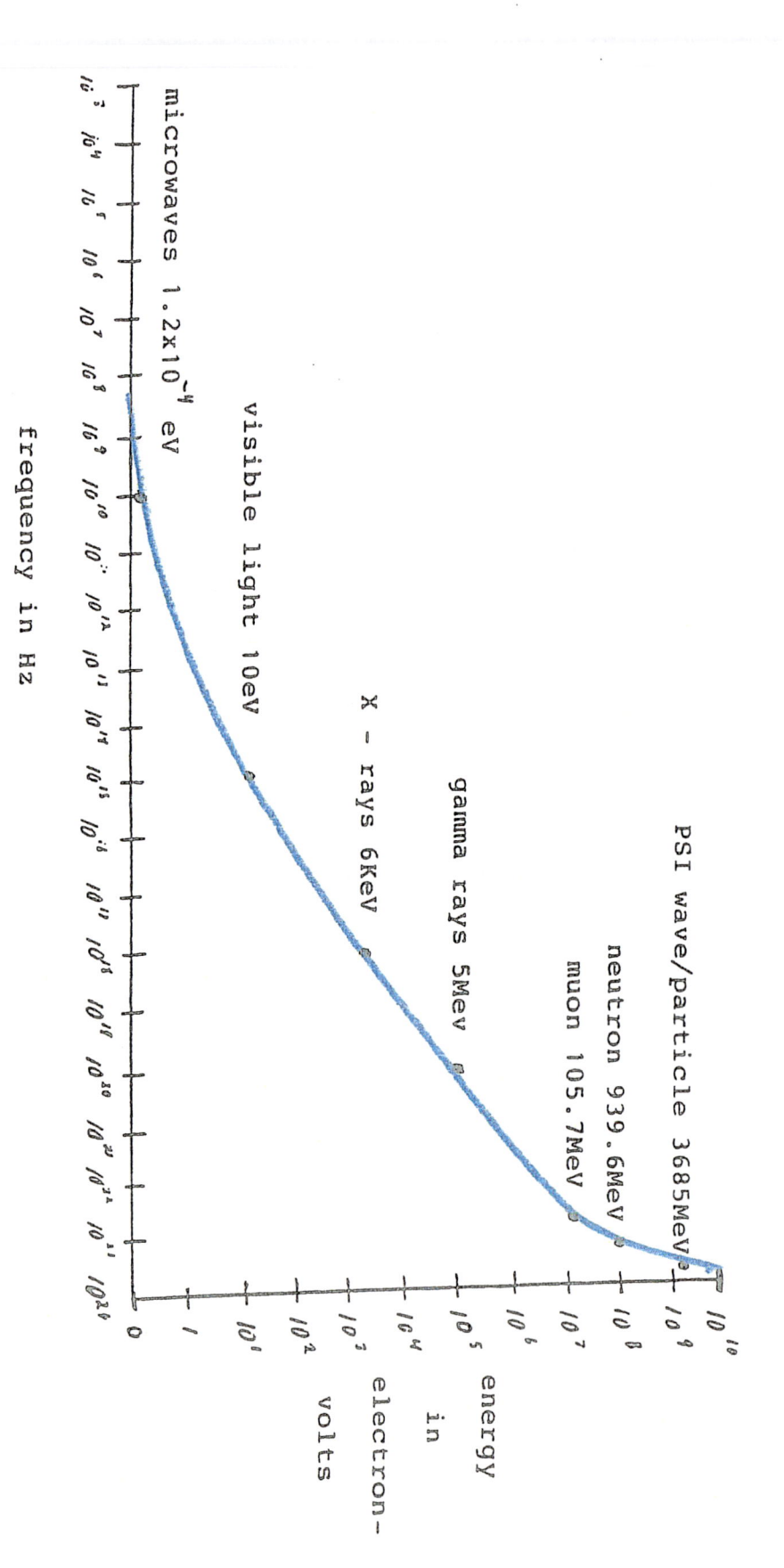

GRAPH OF THE ELECTRO-MAGNETIC RADIATION SPECTRUM

microwaves 1.2×10^{-4} eV

visible light 10eV

X - rays 6keV

gamma rays 5MeV

muon 105.7MeV

neutron 939.6MeV

PSI wave/particle 3685MeV

frequency in Hz

energy in electron-volts

(Continued from page 110)

This is how high-frequency electromagnetic wave/particles, which represent the maximum effects of space-time distortion twice, are able to produce a gravitational field which surrounds them and which is also representative of their total amount of wave energy (mass).

When an electromagnetic wave/particle changes energy states, such as a neutron decaying into a proton, an electron, and a neutrino, the electromagnetic waveform simply shifts to a lower frequency, and in doing so, releases a wave/particle with no electromagnetic characteristics. This wave/particle is called an "electron neutrino" because it is a neutral wave/particle which has a wavelength proportional to the down-shift in frequency that the negative half-wave underwent in order to produce the electron wave/particle (see page 105).

The component parts of atoms have specific energy levels which they tend to maintain, or remain as close to as possible. When an electron absorbs a high-energy photon, for instance, this electron will be raised to a higher energy state (its frequency increases).

In the process, however, the electron becomes unstable and has a tendency to release a photon in order to get rid of the excess energy and go back to its normal energy state. This process is known as photo-emission. The nucleus of the atom is able to undergo the same process, but at much higher energy (frequency) levels (see page 105).

According to this theory, this process is caused by harmonics. All electromagnetic wave/particles tend to seek energy states which are an even harmonic of the neutron frequency, which is the frequency of the initial shock wave in the primordial singularity. Even outside the atom, electrons, protons, and neutrons will last longer without decaying if they are close to an even harmonic of the neutron frequency. The graph on page 111 shows the relationships between harmonic frequencies and regions of stability among various frequency levels.

Surprisingly, the curve at the upper end of the graph appears very much like the upper end of the "time line" as it approaches the primordial singularity (see page 42). This tends to prove many of the relationships in this theory such as: the relationship between relative electromagnetic wave motion and gravity; the relationship between frequency and space-time distortion; and most of all, the relationship between high-energy wave/particles and a "black hole" singularity.

In the next chapter, we will conduct a more in-depth study of the internal characteristics of electromagnetic waves in order to gain a better understanding of how they fit together order to form atoms.

CHAPTER 5

UNDERSTANDING PLANCK'S CONSTANT

When an object is heated it begins to emit electromagnetic waves in four-dimensional space-time within a frequency range between 10^{12} Hz (one with twelve zero's behind it Hz), and 10^{15} Hz (one with fifteen zero's behind it Hz). This frequency range represents the visible light region in the electromagnetic radiation spectrum (see page 93).

Temperature represents the energy of motion of atoms and molecules. Molecules consist of two or more atoms joined together. Atoms and molecules normally vibrate (relative motion) in a crazy, random-like manner which becomes more and more violent as the temperature above absolute zero is increased. The atoms and molecules are still able to remain in their places within the object until the temperature becomes so high that object begins to melt.

As the temperature of the object increases, the electrons within the atoms and molecules begin to produce electromagnetic waves (photons) which are proportional to the amounts of energy they need to release in order to get rid of the additional kinetic (vibrational) energy and return to their normal energy states.

The higher the temperature of the object, the greater the number of electromagnetic waves (light) which are produced by the electrons within the atoms and molecules in the object. As the intensity level of the electromagnetic wave energy, within the visible frequency range of the spectrum, increases, the object begins to glow and becomes brighter and brighter as it continues to get hotter.

Now, imagine that someone bored a large hole, several centimeters deep, into this object, and inserted a length of high-temperature tubing into it (see the illustration on page 117). A 5 cm³ block of a metal known as tungsten can be used in this example because tungsten can withstand very high temperatures without melting. Tungsten is used as filaments in incandescent lightbulbs which essentially serve the same purpose.

As the block of tungsten is heated until it is white-hot, it becomes what is called a blackbody --- a source of light (electromagnetic wave energy) produced entirely by atoms. This experiment should be conducted in a completely dark room and the hot block of tungsten should be hidden from view by a dark shield so that the only light which can be seen is that which is coming through the tubing. This will prevent interference from other light sources.

The light which is coming out of the tubing can be focused onto a prism, and then displayed on a projector screen where the resulting spectrum can be analyzed in more detail. On the projector screen, we will see a continuous spectrum (colors) of electromagnetic wave (light) frequencies ranging from about 10^{12} Hz to about 10^{15} Hz. This process is also quite similar to the way that the sun produces this same type of spectrum, where there is a thick outer layer of heavier elements which is continuously being heated, nearly white-hot, by the nuclear reactions burning in the interior.

This light spectrum, which is the same visible light spectrum which produces the colors that we see in a rainbow, has become known to physicists as the blackbody radiation spectrum (see pages 117 and 118). The graph of this spectrum compares the intensity of the light (shown vertically) with the frequency of the light (shown horizontally) for several different temperature levels of the hot block of tungsten.

In 1900, a German mathematics professor named Max Planck (1859 – 1947) undertook an effort to develop a mathematical formula which would accurately describe the shape of the entire curve on the graph (see page 118) for all of the observed temperature levels and intensities of the electromagnetic waves (light) which are produced within the proximity of the visible frequency range of the blackbody radiation spectrum.

(Continued on page 122)

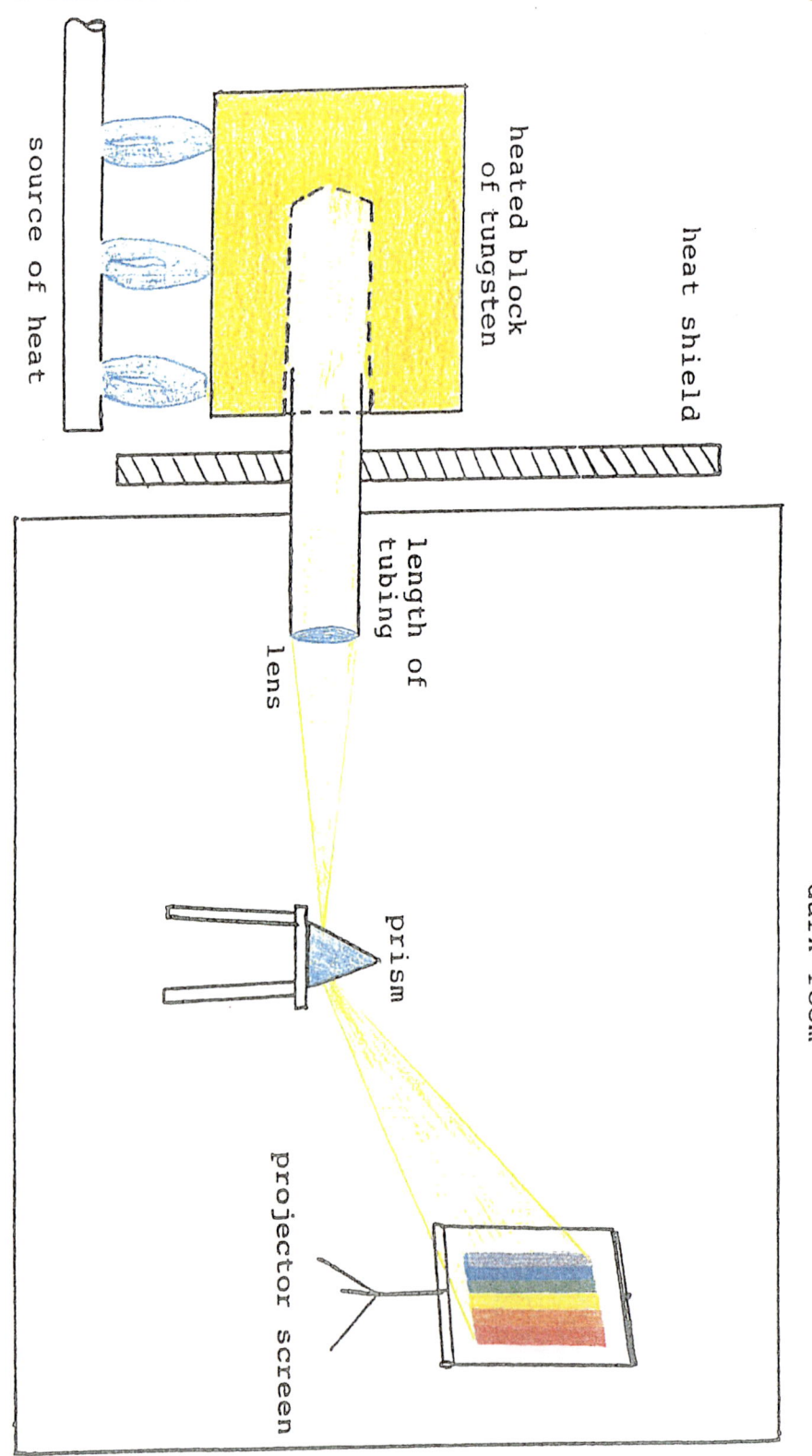

BLACKBODY RADIATION SOURCE

dark room

- source of heat
- heat shield
- heated block of tungsten
- length of tubing
- lens
- prism
- projector screen

GRAPH OF THE BLACKBODY RADIATION SPECTRUM

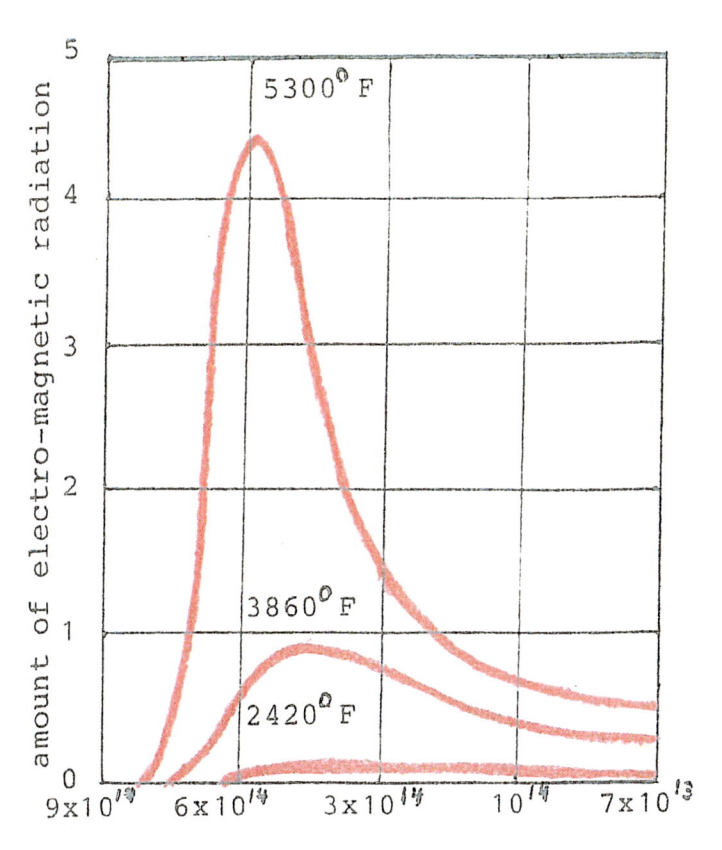

(Graph showing amount of electro-magnetic radiation vs frequency, with curves at 5300°F, 3860°F, and 2420°F. X-axis values: 9x10¹⁹, 6x10¹⁴, 3x10¹⁴, 10¹⁴, 7x10¹³)

The curves on this graph indicate the intensity (amount) of electromagnetic waves given off by atoms as they are heated to various temperature levels. For his formula to work properly a certain "constant" value (h) had to be included. This constant (h), has become known as Planck's constant (h = 6.626,069,57 X 10^{-34} Joule-seconds) (see page 118 and 127).

MAX PLANCK'S FORMULA

$$\text{Intensity} = f(\lambda, T) = \frac{8\pi hc\lambda^{-5}}{e^{hc/\lambda kT} - 1}$$

TERMS:
- c = speed of light
- λ = wavelength in nanometers
- e = elementary charge
- f = frequency
- h = Planck's constant
- k = Boltzmann's constant
- T = temperature in degrees Kelvin

THE SIGNIFICANCE OF PLANCK'S CONSTANT

©2015 by Charles R. Storey

Planck's constant $\quad h = 2\pi c \dfrac{e^2}{\alpha} = $ 6.626,069,57E-34 Joule · Seconds

Reduced Planck's constant $\quad \dfrac{h}{2\pi} = \hbar = c \dfrac{e^2}{\alpha} = $ 1.054,571,725,34E-34 Joule · Seconds

Fine structure constant $\quad \alpha = c \dfrac{e^2}{\hbar} = $ 7.297,352,569,8E-3

Elementary charge squared $\quad e^2 = \hbar \dfrac{\alpha}{c} = $ 2.566,969,732,126E-38 Coulombs

The speed of light $\quad c = \alpha \dfrac{\hbar}{e^2} = $ 299,792,458 meters per second

---

Configurations of the four primary constants which represent unity:

(There are slight inconsistencies in the math)

$$c \dfrac{e^2}{\alpha \hbar} = 1.000,000,000\text{E}9$$

$$e^2 \dfrac{c}{\alpha \hbar} = 9.999,999,955.998\text{E-}26$$

$$\alpha \dfrac{\hbar}{c\, e^2} = 0.999,999,998,690,1\text{E-}9$$

$$\hbar \dfrac{\alpha}{c\, e^2} = 0.999,999,998,69\text{E-}23$$

thus:

$$\dfrac{c\, e^2}{\alpha \hbar} = 1 = \dfrac{\alpha \hbar}{c\, e^2}$$

Four numerically equal expressions of the four primary constants:

$$\frac{e^2}{\hbar} = \alpha\sqrt{\varepsilon_0 \mu_0} = \frac{\alpha}{c}$$

$$\frac{e^2}{\alpha} = \hbar\sqrt{\varepsilon_0 \mu_0} = \frac{\hbar}{c}$$

$$\frac{e^4}{\hbar^2} = \alpha^2 \varepsilon_0 \mu_0 = \frac{\alpha^2}{c^2}$$

$$\frac{e^4}{\alpha^2} = \hbar^2 \varepsilon_0 \mu_0 = \frac{\hbar^2}{c^2}$$

Cross-multiplication produces two numerically equal expressions from the examples above, resulting in the final equation:

$$\alpha\hbar = .000,000,000,000,006,537,188,612,504,934,7 = c e^2$$

Also, the product of $\alpha\hbar$ divided by the product of $c e^2$ yields the above value which represents the quantum jump number for the physical constants related to atoms and electromagnetic waves. This value first appeared in the mathematics regarding these physical constants and appears quite frequently in those particular mathematics and relationships.

Most importantly, when you substitute $\frac{\hbar}{\alpha e^2}$ for " c " in the equations above, the equation vanishes, which strongly suggests that the Permittivity and Permeability values are the means by which the structural orientations within the primordial singularity are represented into our physical reality (our four-dimensional space-time continuum).

The equations above may also be expressed as: $\alpha\hbar = \frac{e^2}{\sqrt{\varepsilon_0\mu_0}}$; and $\alpha^2\hbar^2 = \frac{e^4}{\varepsilon_0\mu_0}$; respectively,

thus: $e^2 = \alpha\hbar\sqrt{\varepsilon_0\mu_0}$; $\alpha = \frac{e^2}{\hbar\sqrt{\varepsilon_0\mu_0}}$; $\hbar = \frac{e^2}{\alpha\sqrt{\varepsilon_0\mu_0}}$; and $e^4 = \alpha^2\hbar^2\varepsilon_0\mu_0$;

$\frac{e^4}{\alpha^2} = \hbar^2\varepsilon_0\mu_0$; $\frac{e^4}{\hbar^2} = \alpha^2\varepsilon_0\mu_0$; $\frac{e^4}{\varepsilon_0} = \alpha^2\hbar^2\mu_0$; $\frac{e^4}{\mu_0} = \alpha^2\hbar^2\varepsilon_0$;

Four numerically equal expressions of the four primary constants:

$$\frac{e^2}{\hbar} = \alpha\sqrt{\varepsilon_0 \mu_0} = \frac{\alpha}{c}$$

$$\frac{e^2}{\alpha} = \hbar\sqrt{\varepsilon_0 \mu_0} = \frac{\hbar}{c}$$

$$\frac{e^4}{\hbar^2} = \alpha^2 \varepsilon_0 \mu_0 = \frac{\alpha^2}{c^2}$$

$$\frac{e^4}{\alpha^2} = \hbar^2 \varepsilon_0 \mu_0 = \frac{\hbar^2}{c^2}$$

Cross-multiplication produces two numerically equal expressions from the examples above, resulting in the final equations:

$$\alpha\hbar = \frac{e^2}{\sqrt{\varepsilon_0 \mu_0}} = c e^2$$

and

$$\alpha^2 \hbar^2 = \frac{e^4}{\varepsilon_0 \mu_0} = c^2 e^4$$

Most importantly, when you substitute $\alpha\frac{\hbar}{e^2}$ for " c " in the equations above, the equation vanishes, which strongly suggests that the Permittivity and Permeability values are the means by which the structural orientations within the primordial singularity are represented into our physical reality (our four-dimensional space-time continuum).

The equations above may also be expressed as: $\alpha\hbar = \frac{e^2}{\sqrt{\varepsilon_0\mu_0}}$; and $\alpha^2\hbar^2 = \frac{e^4}{\varepsilon_0\mu_0}$; respectively,

thus: $e^2 = \alpha\hbar\sqrt{\varepsilon_0\mu_0}$; $\alpha = \frac{e^2}{\hbar\sqrt{\varepsilon_0\mu_0}}$; $\hbar = \frac{e^2}{\alpha\sqrt{\varepsilon_0\mu_0}}$; and $e^4 = \alpha^2\hbar^2\varepsilon_0\mu_0$;

$\frac{e^4}{\alpha^2} = \hbar^2\varepsilon_0\mu_0$; $\frac{e^4}{\hbar^2} = \alpha^2\varepsilon_0\mu_0$; $\frac{e^4}{\varepsilon_0} = \alpha^2\hbar^2\mu_0$; $\frac{e^4}{\mu_0} = \alpha^2\hbar^2\varepsilon_0$;

Permittivity of free space $\varepsilon_0 = \frac{1}{c^2\mu_0} = \frac{e^4}{\alpha^2\hbar^2\mu_0} = 8.854,187,817,62\text{E-}12$ Farads/meter

Permeability of free space $\mu_0 = \frac{1}{c^2\varepsilon_0} = \frac{e^4}{\alpha^2\hbar^2\varepsilon_0} = 1.256,637,061,44\text{E-}7$ Henry's/meter^{-1}

(Continued from page 116)

Max Plank's formula demonstrated that the energy levels of the electromagnetic wave/particles of which atoms consist, and the electromagnetic waves produced by atoms, exist in very small amounts, or "quanta". In other words, a quantum of energy "E" is equal to a certain constant "h" multiplied by the frequency of oscillation " f ", which is described in mathematical terms by the equation: E = h X f.

One of the most surprising results of this historic discovery was that it suggested that electromagnetic wave energy is emitted and absorbed in tiny individual wavelets, called "quanta", each carrying a small amount of energy "E". Thus, the name "Quantum Theory" was derived.

The early discoveries which led to the development of Quantum Theory, indicated that all of the electromagnetic waves, of which atoms consist, and also the electromagnetic waves which are emitted by atoms, all have exactly the same amplitude/angular momentum which is represented by Planck's Constant "h", only their frequency varies.

A closer examination of Planck's constant and the equations on pages 119 and 127 shows how the energy which is carried by an electromagnetic wave actually consists of a combination of three interrelated values: the electric charge (amplitude "e^2") which is represented by ($\alpha \hbar \sqrt{\varepsilon_0 \mu_0}$); the strength of the electromagnetic interaction (coupling force "α") between elementary charged wave/particles ($e^2 / \hbar \sqrt{\varepsilon_0 \mu_0}$); and the angular momentum (ℏ) which is represented by $e^2/\alpha\sqrt{\varepsilon_0\mu_0}$. The quotient of these values is then multiplied by 2π, to represent the total amount of energy (h) which is carried by the wave. This potential energy, which is called an "action" (h), is then multiplied times the total number of individual wavelets which were created by the quantum "event" (frequency) to determine the total amount of energy which is carried by the electromagnetic wave(s) (photon).

Now, according to this theory, Planck's constant " h " also represents the amplitude/angular momentum of the original electromagnetic shock wave in the primordial singularity, at the instant it began expanding and spreading out. Since this original electromagnetic shock-wave in the primordial singularity, is being held in its original "fixed state" within the realm of "basic reality", that is the reason why the amplitude/angular momentum of this initial electromagnetic shock-wave (h) also tends to remain constant, or in a "fixed state".

The fact that all the equations and numerical values associated with Planck's constant (h) are "constants", provides us with even more convincing evidence that the primordial singularity, including the electromagnetic shock-wave which makes up its outer surface, is being held in a "fixed state" by the nature of the void in which it exists.

As the primordial singularity further expanded through a series of subsequent separate existences in the realm of "basic reality" (see page 136), the birth of the four-dimensional space-time energy field caused enormous numbers of separate, relative motion (thermal) identities of this initial shock-wave to develop within the "physical substance" as it continued to expand ("Big Bang").

When the primordial singularity first began to expand, this initial expansion process actually represented the "birth" of four-dimensional space-time, and thus, produced a very large number of separate, relative motion (thermal) identities of the same electromagnetic shock-wave which represents the outer surface of the primordial singularity, all points on the outer surface of our universe, and all points within our universe, simultaneously (see page 20).

All of these separate thermal identities, which are essentially "clones" of this initial electromagnetic shock-wave, represent the extremely large numbers of individual neutrons, as the singularity continued to expand in other "Dimensions" (existences) (see page 136). Most importantly, it is these thermal vibrations (relative motion differences) within all of the separate (neutron) identities of this initial electromagnetic shock-wave, which represents thermal entropy (Brownian motion), or heat, and that is what keeps them all separate.

At that point, during the birth of four-dimensional space-time, the expansion process then became a thermally-driven explosion within our "Dimension" (our universe) and also within the twenty-or-so other "Dimensions" (see page 136) at nearly the same time.

(Continued on page 129)

QUANTUM ELECTROMAGNETIC WAVE RELATIONSHIPS

$$v = v_m \sqrt{1 - \frac{x^2}{A^2}}$$

These drawings are used to provide a two-dimensional representation of the process by which electromagnetic waves (kinetic and potential energy) are related to uniform circular motion.

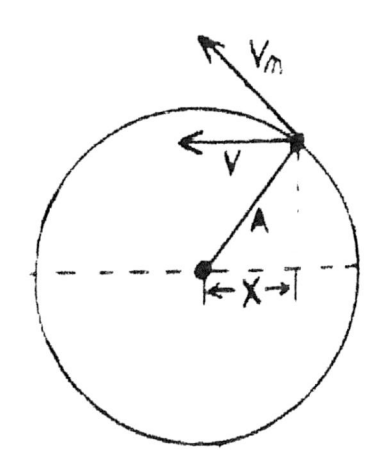

If you can imagine that the point at which lines Vm, V, A, and the vertical dotted line intersect was slowly rotated around the circle at 1 revolution per second, you would see the vertical dotted line moving back and forth along the horizontal dotted line in a manner which represents wave oscillation at a rate of 1 full cycle per second. The point at which Vm, V, and A, intersect can then be speeded up to represent any frequency (rate) of oscillation. "X" represents the actual displacement (compression), both positive or negative, at any point in time, of the medium or substance that these waves (vibrations) are in.

The next drawing can be used to represent harmonic oscillations in any type of medium or substance including the Permittivity (electric charge) and Permeability of free space (magnetic field) of which electromagnetic waves consist. You should always remember, however, that these drawings are a two-dimensional representation of a three-dimensional spherical concept.

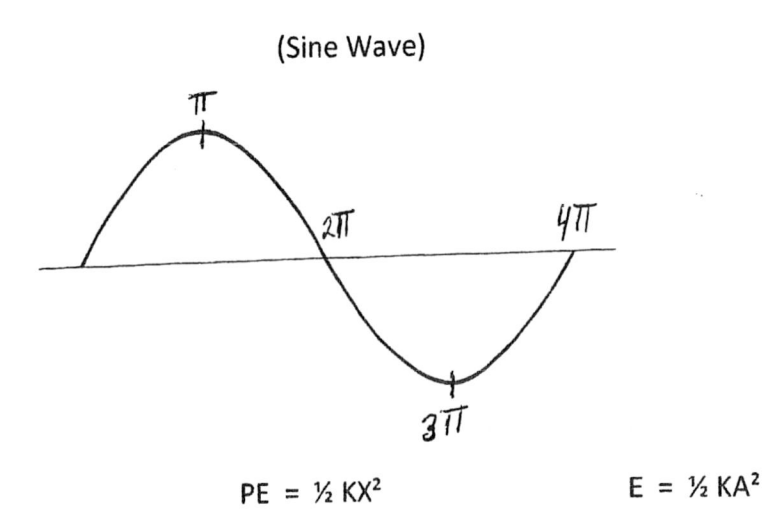

(Sine Wave)

KE = ½ mv²   PE = ½ KX²   E = ½ KA²

As you can see in the drawing on the previous page, the "rate of change" of the medium is the most rapid as it passes the centerline (0) at which point it is all energy in motion (kinetic energy) which is represented by the formula: $KE = \frac{1}{2} mv^2$; or,   kinetic energy (KE) = $\frac{1}{2}$ mass X velocity²

Since kinetic energy is energy in motion, it is usually measured in terms of velocity (v^2). As the wave oscillation in the medium reaches its maximum displacement "A", which represents the amplitude of the wave, it then becomes all potential energy (PE) which represents the maximum amount of physical displacement (distortion) of the spring, or whatever medium the oscillatory motion is in. This gives kinetic and potential energy an inverse relationship to each other in each ¼ wave cycle just like the graphs on page 15. The amount of energy which is held as potential energy in the medium at any given point in time is represented by the formula:

$PE = \frac{1}{2} KX^2$; or,   Potential energy (PE) = $\frac{1}{2}$ spring constant "K" multiplied times the actual displacement² "X²"

Since potential energy represents the energy which is stored in the medium, waiting to be released, it is represented by the actual amount of displacement, or distortion in the medium "X" at any given point in time. This process is the same for all types of waves in all types of physical mediums or substances including the substance which produces our perceptions of electromagnetic waves which is represented by ε_0 and μ_0. The total energy (kinetic and/or potential) (E) at all points along the sine wave is then represented by the formula:

$E = \frac{1}{2} KA^2$; or, $E = \frac{1}{2}$ K (spring constant) X A² (amplitude of the sine wave²)

Changes in the frequency of oscillation, which represent the space-time coordinates of the electromagnetic wave(s), can cause the waveform to be "stretched out" (lower frequency) or "squeezed together" (higher frequency). The harmonic variations in the Permittivity and Permeability values (curve on the graph) will still be representative of 4π regardless of the frequency (time variation) of the electromagnetic wave(s).

Pi (π) (3.141,592,7) is a mathematical constant which is often used to describe our perceptions of circular, spherical, and atomic measurements, including electromagnetic wave motion:

Electromagnetic (harmonic) wave motion	=	4π
The surface area of a sphere with radius "r"	=	$4\pi r^2$
The circumference of a circle with radius "r"	=	$2\pi r$
The area of a circle with radius "r"	=	πr^2

The approximate period " T " of a pendulum of length " L " swinging with small amplitude, "G" is Earth's gravitational acceleration, "m" stands for mass, and "K" is the spring constant.

$$\text{(pendulum)} \quad T \approx 2\pi\sqrt{\frac{L}{G}} \quad ; \qquad \text{(wave motion)} \quad T = 2\pi\sqrt{\frac{m}{K}}$$

The curve on a graph which represents the energy equations is similar to the curve which represents the Lorentz transformation equations (see page 15). This indicates that the same underlying physical process, which is represented by the Permittivity and Permeability values, is producing all of these different "effects":

Momentum of an electromagnetic wave/particle: $\quad p = \dfrac{h}{2\pi\lambda}$

Quantum of angular momentum: $\quad \hbar = \dfrac{h}{2\pi}$

Heisenberg's Uncertainty Principle: $\quad (\Delta x\, \Delta p) \geq \dfrac{\hbar}{2}$

Circular wave number: $\quad K = \dfrac{2\pi}{\lambda}$

Angular frequency in radians per second: $\quad \omega = 2\pi f$

Angular momentum: $\quad L = n\dfrac{h}{2\pi}$

Angular frequency: $\quad \omega = 2\pi v$

The whole number of (electron) wavelengths in the circumference of an electron's orbit:
$$n\lambda = 2\pi r$$

When an electron jumps from level N to N – 1 in an atom, the frequency of the emitted light is equal to:
$$f = \dfrac{v}{2\pi r_N}$$

Below is the mathematical relationship concerning the charge "e" on the electron (-), the proton (+), and Planck's constant "h". The elementary charge is a fundamental physical constant which is one of the main components of an electromagnetic wave. Planck's constant represents the total amount of potential energy which is contained in the amplitude/angular momentum of a single individual electromagnetic wavelet and also the initial electromagnetic shock-wave in the primordial singularity:

$$e^4 = \alpha^2 \hbar^2 \varepsilon_0 \mu_0 \; ; \qquad h = 2\pi \dfrac{e^2}{\alpha\sqrt{\varepsilon_0 \mu_0}} \; ; \qquad \hbar = \dfrac{e^2}{\alpha\sqrt{\varepsilon_0 \mu_0}} \; ; \qquad \alpha = \dfrac{e^2}{\hbar\sqrt{\varepsilon_0 \mu_0}} \; ;$$

$$\frac{e^4}{\alpha^2} = \hbar^2 \varepsilon_0 \mu_0 \; ; \quad \frac{e^4}{\hbar^2} = \alpha^2 \varepsilon_0 \mu_0 \; ; \quad \frac{e^4}{\varepsilon_0} = \alpha^2 \hbar^2 \mu_0 \; ; \quad \frac{e^4}{\mu_0} = \alpha^2 \hbar^2 \varepsilon_0$$

Permittivity of free space $\varepsilon_0 = \dfrac{1}{c^2 \mu_0} = \dfrac{e^4}{\alpha^2 \hbar^2 \mu_0} = 8.854,187,817,62E-12$ Farads/meter

Permeability of free space $\mu_0 = \dfrac{1}{c^2 \varepsilon_0} = \dfrac{e^4}{\alpha^2 \hbar^2 \varepsilon_0} = 1.256,637,061,44E-7$ Henry's/meter^{-1}

The constants: 2π, ε_0, μ_0, α, c, e, h, and $\hbar$ are frequently used in quantum physics for electromagnetic wave/particle relationships, descriptions, and interactions, thus, revealing the underlying physical make-up of the basic electromagnetic waveform.

LEGEND

α = Fine structure constant — It represents the strength of the electromagnetic interaction between elementary charged particles.
A = Maximum displacement from the equilibrium position
c = speed of light
e = elementary charge = 1.602,176,560,846,5E-19 Coulombs
e^2 = 2.566,969,732,126E-38 Joule · Seconds
E = energy
h = Planck's constant = 6.626,069,57E-34 Joule · Seconds
$\hbar$ = Reduced Planck's constant ($\hbar = h / 2\pi$)
k = effective force constant
K = circular wave number
K = spring constant
KE = kinetic energy
λ = wavelength
m = mass
n = number, as in number of waves, or, number of wave/particles
N = electron shells (energy levels) in an atom
PE = potential energy
r = radius
v = velocity
v_m = maximum Velocity
X = Displacement from the equilibrium position --- the center point is the equilibrium position
ω = angular frequency
Δp = uncertainty in the momentum measurement of a wave/particle
Δx = uncertainty in the position measurement of a wave/particle

(Continued from page 124)

It was this thermally-driven expansion process which accelerated all of the newly produced neutrons (within all of the "Dimensions") to such a very high outward relative velocity, even though the primordial singularity itself was being held, or "remembered" in its original "fixed state" within our "Dimension", and also, in all of the other "Dimensions", by the nature of the realm of "basic reality".

That is most likely where the singularity which represents our "Dimension" (our observable universe) is actually physically located --- it was originally divided up among all of the individual neutrons within our universe. In other words, the wave/particle existences of all of the neutrons within our early universe, cumulatively, represent the primordial singularity in our "Dimension" (universe).

This initial physical expansion process of the "physical substance" itself, is what produced all of the relativistic, thermal/relative motion/location differences which gives all the neutron "clones" their different identities and keeps them all separate from each other.

Additional evidence supporting this hypothesis may be found in the results of still-ongoing experimentation with Bose – Einstein Condensates. In 1924 – 1925, Satyendra Nath Bose (1894 – 1974) and Albert Einstein predicted that matter (atoms and molecules), which were cooled to temperatures near absolute zero, would exhibit some very interesting properties.

During the course of these experiments it was discovered that some super-cooled substances became unstable and would collapse beyond detection with the exception of only a few molecules. In other instances, super-cooled substances collapsed beyond detection and then exploded (thermally), after which, only half of the original atoms, which were confined within the experimental device, could be detected in the remaining cold material or in the expanding gas cloud. During other experiments, large numbers of atoms and molecules disappeared from the experiment altogether and could not be found.

The formation of Bose-Einstein condensates appear to be quite similar to the predicted formation and functioning of a black hole. Some researchers are even considering using Bose–Einstein Condensates for modeling black holes and their related phenomena in laboratory settings. This interesting line of research is providing us with even more convincing evidence that it is the thermal/relative motion differences between the individual neutron identities which is keeping them all separate.

Around that same time period (1924 – 1925), two other physicists, Erwin Schrödinger (1887 – 1961) and Werner Heisenberg (1901 – 1976) independently developed two different mathematical approaches to the establishment of a new type of quantum theory which eventually became known as Quantum Mechanics.

Werner Heisenberg used matrices to describe quantum phenomena, which was undoubtedly a very important contribution. Erwin Schrödinger's approach was to develop an equation which would enable us to calculate the probability amplitude, or "wave function" (Ψ) of matter waves (electromagnetic waves above 10^{20} Hz) at any point in space and time.

This "wave function" is representative of the different degrees of freedom for the various physical properties (quantum states) such as: position, momentum, spin, angular momentum, polarization, etc., regarding all of the electromagnetic wave/particles throughout our universe.

Where (Ψ^2) is higher, there is a higher probability of finding the particle existence of an electromagnetic wave/particle at a particular location in space and time. Where (Ψ^2) is lower, the probability of finding the particle existence of an electromagnetic wave/particle at a certain point in space and time is lower.

The position in four-dimensional space-time for any electromagnetic wave/particle can also be written as: $\Psi(R,T)$; where " R " is the position vector in three-dimensional space; and " T " is time. The familiar Cartesian Coordinate System can be used for " R " in this particular example by plotting the location on the X - axis, Y - axis, and Z – axis.

In the 1960's, it was discovered that light wave/particles (photons) could be made to react to certain situations in a very strange way. Two separate beams of light which were emitted from the same source (atoms) and then passed through a beam-splitter (a partially silvered mirror) and directed in different directions, could be made to react (change) as if they were still together, or still in the same place. If something was done to one beam, it would affect the other beam, and vice versa.

The distance between light beams was increased to the point where there was no way anything could "warn" the other beam of light as to what was going to happen, unless there were some type of signals that could travel faster than light. It was also discovered that beams of electrons could actually "predict" the configuration of a measuring device even before they entered it.

Many other experiments, have demonstrated that certain quantum processes involving electromagnetic wave/particles which are produced by atoms, are intimately related and instantaneous across large distances of space. This phenomenon has become known as "quantum entanglement". If a measurement is made on one of the wave/particles, the other(s) will "know" instantly what measurement has been made, what the outcome of that measurement was, and then react to it accordingly.

The results of quantum experiments such as these, led to the development of the "Space Chart" on page 20 and the "Time Chart" on page 42. By comparing these two charts, you can see that all quantum events have a "point of origin", with the point of "basic reality" being the "point of origin" of our entire universe. The space and time relationship of any quantum event can be plotted on the "Space Chart" by superimposing the center point ("point of origin") of its "Sphere of Influence" (space characteristics) over its "point of origin" (time characteristics) on the "Time Chart".

The "Sphere of Influence" on the "Space Chart" refers to the fact that a wave in the four-dimensional space-time energy field appears to us like an expanding sphere, moving outward in all directions at the speed of light from the point in space where the event originated ("point of origin").

The "point of origin" on the "Time Chart" simply refers to the point in time between the "Big Bang" and now, when the quantum event occurred. This will give you the space and time relationship of any quantum event, such as the creation of an electromagnetic wave/particle, in the four-dimensional space-time energy field.

Now the geometric dilemma that we are facing in this universe is that the "effects" of the four-dimensional space-time energy field, which are represented by the Permittivity and Permeability of free space, make it look to us like the "point of origin" of the "Sphere of Influence" exists simultaneously at all points on its surface and also at all points within the sphere (space), simultaneously. Also, that the "Sphere of Influence" of the quantum event is expanding outward in all directions at a certain rate (time) which, of course, is the speed of light.

ε_0 the Permittivity of free space: 8.854,187,817,6E-12 Farads per meter.

μ_0 the Permeability of free space: 1.256,637,061,44E-7 Henry's per meter^{-1} (The $^{-1}$ exponent means that there is an inverse relationship between ε_0 and μ_0).

Our universe, as it has expanded and spread out until the present time (our four-dimensional space-time continuum), is representative of all the cumulative wave functions (Ψ) for all of the original neutrons before they began to decay into protons and electrons. As a result, the median strength of the Permittivity and Permeability values throughout our universe are representative of this cumulative wave function (Ψ) for all of the original neutrons.

A neutron, which represents the initial electromagnetic shock-wave in the primordial singularity, has a "Sphere of Influence" or wave function (Ψ) which **is** the size of the outer reaches of our universe. In other words, the "Sphere of Influence" of all the neutrons **are** the universe.

Thus, the wave function of the neutrons (Ψ) simultaneously represents both the amplitude/angular momentum of all the neutron electromagnetic wave/particles at any point within our universe, and also the "probability amplitude" of finding the wave/particle existence of the neutrons at any point within, or, on the outer surface of our universe.

Furthermore, every electromagnetic wave/particle which was created during the time of the initial shock-wave in the primordial singularity, has a "Sphere of Influence" (wave function), which is a small part of our whole universe, and contributes to the collective strength of the Permittivity and Permeability values, right on down to the "Big Bang" background blackbody radiation which persists within our universe until the present-day.

As a result of this situation, the particle existences of atoms and their component parts can exist at any time or at any place within our universe, and, consequently, experience "quantum entanglements", "quantum teleportation", and "tunneling" through barriers. This is why atoms, and their component parts, can "predict" the configuration of a measuring device before they enter it.

This is also what causes all the crazy quantum "jumping" from one energy state to another and from one place to another as observed in Quantum Physics Laboratories. Consequently, the "Sphere of Influence" (wave function) of all the neutrons, including the component parts of the neutrons (electrons and protons), and also the earliest waves (photons) emitted by them, are all represented by the diameter of our universe (see page 20).

The Permittivity and Permeability values of our universal four-dimensional space-time energy field are representative of all the wave functions, which include the cumulative probability amplitudes (capacitance/electric charge) and the (inductance/magnetic field) at all points within the "Sphere of Influence" (our universe) of all the neutrons.

So, according to this theory, Planck's constant " h " represents the amplitude/angular momentum of the particle existence of the initial electromagnetic shock-wave in the primordial singularity at the instant before it started expanding.

(Continued on page 137)

EXPANSION OF THE PHYSICAL SUBSTANCE

© 1995 by Charles R. Storey

Empty Black Void

LUMINIFEROUS ETHER (CLOUD) WHICH PRODUCES OUR UNIVERSE THAT WE SEE (WHERE IT IS NOW)

Empty Black Void

$S = 4\pi R^2$ THIS EQUATION DESCRIBES THE SPHERICAL SHAPE OF THIS (CLOUD) OF LUMINIFEROUS ETHER SUBSTANCE

$\dfrac{1}{R^2}$ LAW

Describes the physical expansion of the luminiferous ether (cloud) which produced the four-dimensional space-time power field.

Fitzgerald - Lorentz transformation equations:

$$T^1 = \dfrac{T}{\sqrt{1-\left(\dfrac{v}{c}\right)^2}}$$

$$L^1 = L\sqrt{1-\left(\dfrac{v}{c}\right)^2}$$

These equations taken together, actually describe distortions in the four-dimensional space-time power field which produces our perceptions of: space, time, relative motion, acceleration and gravity.

The four-dimensional space-time power field which produces our perceptions of time and space is caused by the tremendous pressure difference of the luminiferous ether between its starting point (primordial fireball singularity) and its expanded form (where it is now).

OUR DIMENSION (EXISTENCE)

$c = \dfrac{1}{\sqrt{\varepsilon_0 \mu_0}}$

Starting point

The black dotted lines represent different (dimensions) existences of the luminiferous ether cloud. These separate existences are caused by the strange nature of the empty black void in which our universe exists. All of these separate existences exist as concentric spheres in all of their respective sizes and pressures simultaneously, as if they are "stuck" within the empty black void. They were drawn here in this manner to show that all of the different "existences" are separate from the others.

This equation represents the pressure/density of the luminiferous ether in the primordial fireball singularity which represents our dimension (existence).

(Continued from page 135)

The wave function (Ψ) represents the amplitude/momentum of the wave existence of that same electromagnetic shock-wave as it has expanded and spread out to the size it is today (the size of our universe).

The expansion process of the "physical substance", which produces our perceptions of our expanding universe, is still continuing and is causing time to pass as we know it (see page 15). Consequently, as our universe continues to expand, the "Sphere of Influence" of all the original neutrons and their progeny (electrons, protons, and the photons emitted by them) continues to expand also.

As seen on the "Space Chart" (see page 20), the center point, or "point of origin" of an electromagnetic wave/particle, represents all of the points on the surface of its "Sphere of Influence" as well as all of the points within its "Sphere of Influence", simultaneously (Quantum Theory).

Now, let us consider a photon which has been emitted from a wave/particle that has the primordial singularity as its point of origin. This photon will "inherit" the "Sphere of Influence" (wave function) of its parent electromagnetic wave/particle, in addition to its own, newly created, "Sphere of Influence". This situation is what is causing all the faster-than-light responses, which have been observed during quantum experiments.

Shortly after the neutron wave/particles came into existence, many of them began to decay (break apart) to form protons and electrons. The half-life of a free neutron is approximately 15 minutes. This means that if you filled a container with free neutrons, within fifteen minutes, half of them would decay (break apart) into protons and electrons.

Some of these protons and electrons will then recombine to form hydrogen atoms which consist of the proton in the nucleus and the electron in a resonant wave-like orbit around it. This process is a commonly-observed phenomenon in particle physics laboratories.

Within our early universe, some of the free neutrons, were then "captured" in the nuclei of the hydrogen atoms to form helium, which consists of a neutron and a proton in the nucleus, with one orbiting electron. These two types of atoms are very stable and will last for a very long time without decaying. A proton decay has never been observed, and a neutron becomes stable as long as it is held in the nucleus of an atom.

Large clouds of this newly-formed hydrogen and helium gas condensed and became highly concentrated under gravitational pressure to start the nuclear proton-proton reaction (burning process) within the earliest stars. The nuclear reactions within the stars are what enabled the heavier elements (more complex atoms), and molecules to form.

Some of the larger stars, including the larger stars in binary systems, consumed their hydrogen and helium fuel more rapidly than others and exploded as supernovae which eventually led to the formation of solar systems and planets like ours.

However, all of these more-complex atoms and molecules, including the photons produced by them, still retain their respective proportions of this cumulative universal wave function, which is currently being represented by the median Permittivity and Permeability values throughout our universe (Dimension).

Electromagnetic wave/particles "maintain" their existence back at their point of origin on the "Time Chart" (see page 68) partly because of their electromagnetic characteristics, and partly due to velocity-related space-time distortion. The electromagnetic characteristics and also the velocity-related space-time distortion, which are both represented by the Permittivity and Permeability values, are instantaneous across space and time between the primordial singularity which still exists in the realm of "basic reality", and "where we are now".

So if you do something to one of these wave/particles, such as observe it, or if it is "captured" by another electromagnetic wave/particle, its existence is affected all the way back to its "point of origin" and all across its "Sphere of Influence" (wave function collapse).

This is why these high-frequency electromagnetic wave/particles have such a strange existence, so that the more you know about their location (space coordinates), the less you can know about their momentum (time coordinates). This is another example of the inverse relationship between the space and time coordinates of an electromagnetic wave/particle within our space-time continuum (see page 15). This is why you can only tell where they are and what their momentum is, just while you are actually observing them (Uncertainty Principle) (see page 127).

The fundamental interactions between electromagnetic waves/particles, which enable atoms and molecules to exist, are certainly a fascinating aspect of quantum physics and Quantum Mechanics.

One of the most important aspects of these basic principles deals with the discoveries regarding the way in which electrons form a "cloud of charge" around the nucleus of an atom. As we have seen previously, electrons, which represent the negative half-wave of the initial shock-wave in the primordial singularity, have a frequency of around 10^{21} Hz.

Atoms and molecules function according to the principle of electromagnetic wave resonance, so that the diameter of each of the various electron "shells" in an atom, are determined by wave harmonics, or three-dimensional standing waves. Only electrons with an even number of wavelengths (frequency) which represent the precise circumference of the electron orbits are able to occupy the various electron energy shells (orbits) within an atom.

Other quantum values (quantum states) include the polarization of the electric charge and magnetic field, transverse or longitudinal wave polarization, orbital angular momentum, magnetic angular momentum, and spin (intrinsic angular momentum). Only electrons with different quantum values (quantum states) are able to share the various electron shells. No two electrons in the same quantum state are allowed to occupy the various electron shells within the same atom (Pauli Exclusion Principle).

The spin value is the quantum value of an electromagnetic wave/particle which represents the amount of energy consumed by the angular momentum of the physical substance as it "changed" from its lower pressure (-) state, to its higher pressure (+) state. This intrinsic angular momentum energy value is also held in a "fixed state" within the realm of "basic reality" and contributes to the physical, or "quantum", characteristics of an electromagnetic wave/particle within an atom.

The electromagnetic wave/particles in the nucleus of an atom (protons and neutrons) must also remain within specific wave orientations and energy levels in order to maintain stability.

There are many more additional details regarding electromagnetic wave orientations within atoms and molecules to be found in Quantum Mechanics. Several books have been written on the subject. However, it should be obvious to the reader by now that atoms and molecules, which are the fundamental components of our universe as a whole, are comprised of electromagnetic waves at resonant frequencies, and that mass (atoms) consists of electromagnetic wave energy.

One of the unusual sets of circumstances which needs to be further clarified here, is that a neutron, which represents the initial shock-wave in the primordial singularity, is not a uniform electromagnetic wave (see drawing below). The neutron electromagnetic wave is "asymmetrical" which means that its positive half of the wave is not exactly the same as its negative half of the wave.

The proton which represents the positive, or (high pressure) half of the wave, also represents the outer (spherical) surface of our universe. Since the proton has "nothing" in front of it (the realm of "basic reality"), that situation may account for its extremely slow rate of decay. Also, the wavelength of a proton is approximately 1,823 times shorter than the wavelength of an electron, which represents the negative half of the same electromagnetic wave.

This accounts for the very large difference between the mass of a proton and the mass of an electron, and also that the mass of a proton, plus the mass of an electron, do not quite equal the mass of a neutron. A cross-section of the electromagnetic wave existence of a neutron would actually look something like this:

(two-dimensional view)

The potential-energy amplitude of the neutron, which is represented by Planck's constant, is approximately equal to 6.625×10^{-34} Joule-seconds.

The electric charge on the proton, or positive half of the wave, is approximately equal to $+ 1.602 \times 10^{-19}$

Coulombs, and the electric charge on the electron, or negative half of the wave, is approximately equal to − 1.602 X 10^{-19} Coulombs. The measuring system which enables us to quantify the electric charge on a proton or an electron, can also enable us to quantify the Permittivity value for each point in space and can be measured in Coulombs² / Newton · meter².

One of the main characteristics of gravity that applies to quantum physics is the way in which high-frequency electromagnetic wave/particles create a gravitational field. As we have seen earlier, matter consists of atoms which consist of protons, neutrons and electrons. Each of these high-frequency (above 10^{20} Hz) wave-particles creates its own small, velocity-related space-time distortion manifold, which can be considered to be a miniature black hole, and that produces a corresponding distortion in the Permittivity and Permeability values in the space surrounding an atom.

If you multiply the amplitude/angular momentum (Planck's Constant " h ") by the frequency of the electromagnetic wave/particle, this will give you the average amount of velocity-related space-time distortion created by the wave-particle. Since gravity "is" velocity-related space-time distortion, the sum of the average amounts of velocity-related space-time distortion created by each wave/particle within an atom determines the relative strength of the gravitational field which surrounds the whole atom.

In other words, the sum of the relative strengths of the miniature gravitational manifolds at the respective distances according to the $\frac{1}{r^2}$ law, determines the total amount of velocity-related space-time distortion (gravity) surrounding the whole object.

If you take another look at the gravitational manifolds on page 51, you can see that the miniature "black hole" manifolds, which are representative of the diameter of the electromagnetic wave/particles (10^{-15} cm), would also be very, very small because atoms actually consist of mostly empty space. Consequently, the event horizon, which is the radius at which the space-time distortion (gravity) really becomes significant, would be extremely close to the "fuzzy" surface of the wave/particle.

This would also lead us to believe that the strong nuclear force, which has the ability to overcome repulsive electric charges, is just an ordinary gravitational attraction which does not become really significant until another massive electromagnetic wave/particle, with its own miniature black hole gravitational manifold, gets within a very, very close proximity to the surface of the wave/particle.

Due to the fact that the total amount of velocity-related space-time distortion (gravity) surrounding each atom (measured in terms of acceleration) at even small respective distances is very, very small, this is why it takes an object with a very large number of atoms, such as the Earth, to produce a sizeable gravitational field.

The four-dimensional space-time energy field (Permittivity and Permeability values) also represents the "Time Lock" which keeps track of where and when in space and time everything is located within our universe throughout the ages. It keeps track of all the gravitational attractions (accelerations) including the locations and states of relative motion of all the bodies (objects) including all of the atoms, molecules, electrons, protons, neutrons, photons, etc., within our universe.

The four-dimensional space-time energy field keeps track of where all the atoms in our universe are located (space) at any point in time (time), and it also keeps track of all the thermal relative motions of all the atoms and their component parts including electric charges and magnetic fields.

Another characteristic of the primordial singularity as it exists in the realm of "basic reality" is its ability to record every quantum electromagnetic event which has ever taken place in the order in which it has occurred, from the time that the primordial singularity began expanding outward, and in all of its separate existences up until now (see page 136).

Since the interaction between the physical substance in the primordial singularity and the realm of "basic reality" has a tendency to hold all the "pressure related" distortions and all the expansion-related (velocity-related) distortions, of which quantum events consist -- in a "fixed state" -- this would lead us to conclude that every single quantum (electromagnetic) event which has occurred throughout the entire history of the expansion process of our universe, would all remain "fixed" within each of the various separate existences of the primordial singularity right up until the present time.

This would also indicate that if you could develop a satisfactory system for distorting the four-dimensional space-time energy field, which is represented by the

Permittivity and Permeability values, you would be able to travel back in time, or into the future, and witness everything occurring just the way you do in the present.

So far, we have seen how events on an astronomical scale are related to events at the atomic level. Also, we can see how the strange characteristics of our four-dimensional space-time energy field, which is represented by the Permittivity and Permeability values, combine to form the world in which we live. Now, in the next chapter, we will take a look at how the "effects" of four-dimensional space-time may be used to benefit us in ways we have never even dreamed of --- until now.

CHAPTER 6

THE FUTURE OF SPACE EXPLORATION

In previous chapters, we discussed the ways in which electromagnetic fields are produced by distortions in the four-dimensional space-time energy field which produces our perceptions of time, space, energy, relative motion, and gravity. Now, let's see how we can use electromagnetic fields to distort the effects of four-dimensional space-time.

Since we have seen how the Permittivity and Permeability of free space work together to determine the relative motions of objects, the passage of time, our perceptions of space (size), and acceleration which is perceived by us mostly as gravity, then, at least theoretically, we should be able to build an electronic device which has the ability to manipulate the Permittivity and Permeability values of free space in such a manner as to provide us with a new type of space travel.

A word of caution to all individuals who would attempt to conduct any type of experimentation concerning four-dimensional space-time distortion is that they need to be aware of the fact that these types of experiments should only be carried out by professional, highly trained, scientific personnel taking every possible safety precaution which can be devised. If any mistakes are made, or if anything should go wrong during the course of such an experiment, disaster will almost certainly result.

With these safety precautions in mind, you could start the sequence of experiments by constructing a cylindrical metal enclosure approximately 45 centimeters in diameter, and approximately 15 centimeters deep, preferably made of copper with a brite-finish on both the inside and outside surfaces (see pages 149 and 150). The edges can be silver-soldered together and the inside of the cylindrical container can be coated with a silver-plating compound such as "Cool Amp", which would make the inside surfaces very highly reflective to electromagnetic waves. This type of cylindrical metal enclosure is referred to as a "resonant cavity".

(Continued on page 151)

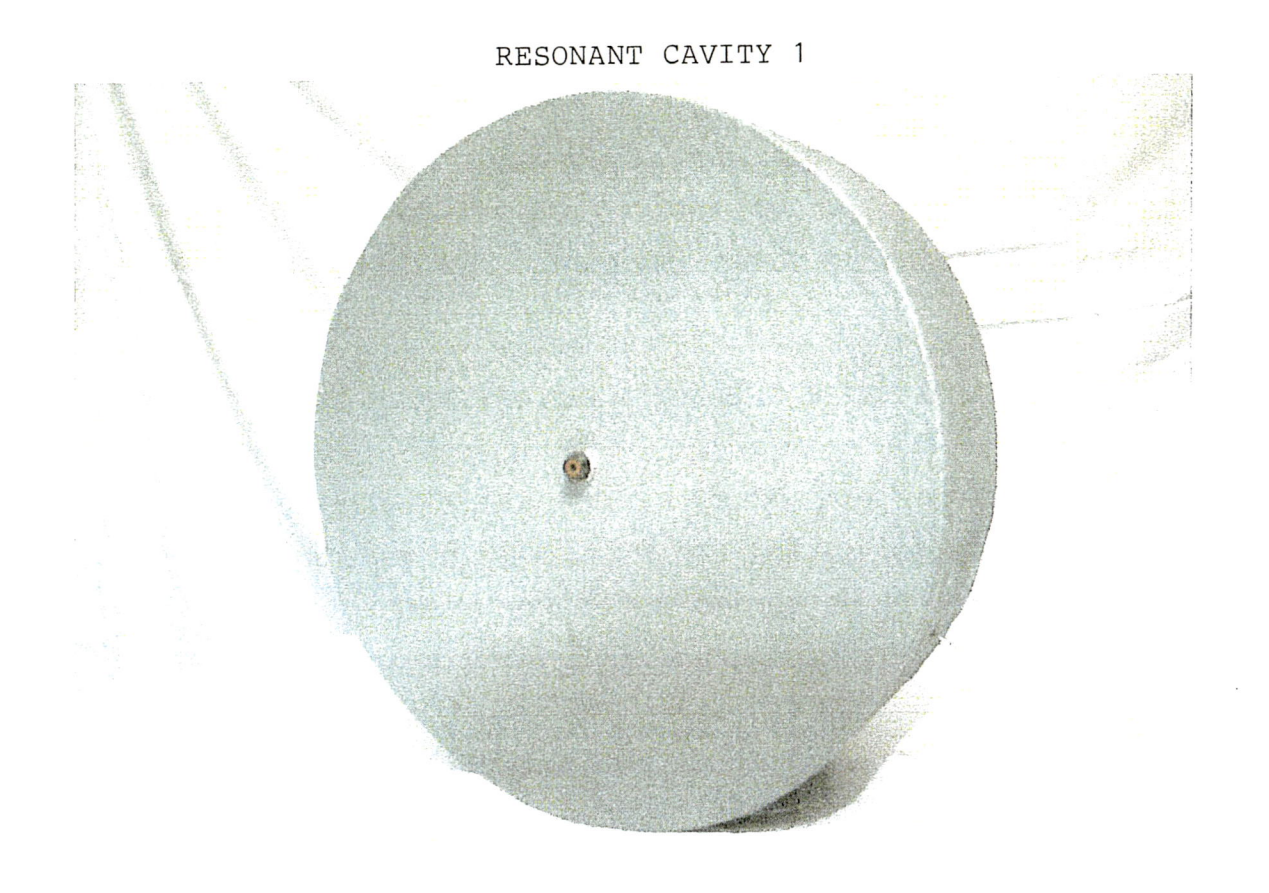

RESONANT CAVITY 1

RESONANT CAVITY 2

The next step is to cut a 10 centimeter diameter hole, centered (centric), in one end of the cylinder, which can then be covered by a tight-fitting, access door, which is also silver-plated on the inside. Then, a small hole is bored, centric, in the other end of the cylinder, through which a coaxial cable connector can be installed and "flush-mounted" with the inside of the cylinder.

A small, L – shaped antenna can be constructed which is designed to protrude 1 centimeter out from the coaxial cable connector in the center of the back wall of the chamber and which is then bent 90° so that it runs parallel to the surface of the back wall of the cylinder for a distance of approximately three centimeters. This L – shaped antenna can be easily constructed from #12 AWG bare copper wire, which should then be connected to the inside conductor of the coaxial cable connector and also silver-plated for greater efficiency (see pages 149 and 150).

A wide frequency-range spectrum analyzer can be connected to the outside surface by a wire connector soldered to the rounded side of the "resonant cavity", or "resonator" (see lower photograph on page 149 and 150), to monitor the intensity of the electromagnetic fields inside the unit and to help determine the optimum frequency for operating the system. The optimum resonant frequencies can then be determined by experiment and the length of the antenna wire can be adjusted to provide maximum efficiency.

The outside of the cylindrical container may also be coated with a thin layer of bismuth, which can be applied with heat, like solder, and which is somewhat reflective to the magnetic portion of the waves. This helps to keep the electric and the magnetic wave-crests in-phase and greatly improves the overall efficiency of the device.

Now, according to this theory, you will have constructed a device which can be brought to resonance and will be capable of producing various amounts of space-time distortion in the space within the unit. The next step is to use this device which, we shall call a "resonator", to conduct a series of experiments designed to optimize the physical dimensions (size), frequencies, voltages, antenna design, materials, signal-generating equipment, etc. The most effective types of test equipment will need to be selected which will be able to measure and record the various space-time distortion effects within a high degree of precision.

The basic scientific principle upon which this device operates is known as electromagnetic wave resonance. For a simple comparison, the operation of this device can be compared to filling a flat-bottomed, in-ground swimming pool, 15 meters wide by 30 meters long, half full of water. The water represents the medium which conducts the waves and which also can be roughly compared to the Permittivity (capacitance) and Permeability (inductance) of free space in this particular example.

You could then use a wave-making machine which would be able to make large waves in the water, evenly, all the way across the pool, resulting in a uniform (resonant) wave motion back and forth between the ends of the pool. As the wave-maker produces the wave motion back and forth and the waves continue to increase in amplitude, the wave crests are reflected back from the ends of the pool and some of the water may soon splash out over the ends. The troughs of the waves, however, would reach deeper and deeper toward the bottom of the pool.

When the resonant wave motion becomes sufficiently violent so that the troughs of the waves reach all the way down to the bottom of the pool, the medium (the water) would be split, or divided, and cease to have any "effect" in those particular areas at the bottom of the pool. Now this is exactly what will happen to the Permittivity and Permeability values inside the resonant cavity as the electromagnetic waves are brought to resonance and begin to form "standing waves".

The next thing which we need to be concerned with at this point, now, is efficiency. When an electromagnetic wave is reflected from an ordinary metal surface such as a waveguide in a radar system, the electric charge portion of the wave is all that is actually being reflected. The magnetic field portion of the wave passes right on through the metal surface of the reflector, and then it has to be "pulled back in" by the electric charge portion of the wave which ends up causing the waveform to become distorted.

If the electromagnetic wave is to be reflected back and forth a number of times in order to dramatically increase its amplitude, then this situation can produce undesirable distortions in the electromagnetic waves and prevent the system from operating efficiently.

That type of unsynchronized electromagnetic wave reflection is sufficient in most cases, for applications within our telecommunications industries. However, in this particular situation, both the electric charge and the magnetic field need to be reflected equally, and simultaneously, in order to maintain the integrity of the standing waves and to keep the electric charge and magnetic field portions of the waves properly in-phase.

Research efforts to determine the best method for reflecting magnetic fields have led to the use of bismuth for this promising line of research. Bismuth has some peculiar electromagnetic qualities, and by the process of layering various other metals with thin layers of bismuth, sufficient reflection of the magnetic fields can be obtained to keep the electric charges and magnetic fields in-phase and thus, reflect (resonate) the electromagnetic waves efficiently.

Now, if this "resonator" could be constructed as previously described, with the addition of the bismuth and alternately layered metals, then we would be enabled to conduct much more sophisticated, in-depth, research into the quantum space-time distortion effects of electromagnetic wave resonance.

Again, these types of experiments should only be conducted by highly trained, professionals working under closely controlled laboratory conditions. In addition to the space-time distortion itself, there may be other hazards that are not readily apparent including internal heating and electromagnetic radiation exposure which could be harmful to the experimenters if they are not properly protected.

Now, imagine that an enterprising group of physicists were able to construct a large "resonator" using the same technology as the one which was previously described, say, fifty meters in diameter. This type of resonant cavity does not necessarily have to be cylindrical, either. Various sizes and shapes can be experimented with in order to provide the optimum standing wave-form densities and shapes.

Next, they could build a spacecraft which includes this type of "resonator". If this spacecraft were outfitted with the necessary power sources, landing capabilities, life support systems, control equipment, signal generating equipment, etc. These physicists would be equipped with a vastly superior system of space travel and exploration than that which we are currently using.

This type of space travel would be a wonderful opportunity for us to extend our space exploration efforts beyond our solar system, providing that all the necessary precautions are taken. When considering the obvious hazards which would be associated with this form of space travel, it can be compared to the early pilots who flew the first airplanes.

With very little understanding of the many hazards associated with aviation, these brave individuals experimented with often poorly designed, ill-equipped, flying machines, and risked their lives to promote aviation technology.

Of course, the early pilots encountered numerous difficulties, but progress continued regardless of the challenges. Today, when modern aircraft are flown properly and safely, they are a wonderful benefit to all of us.

In light of the obvious safety risks, the initial research and development with this new technology would probably be best conducted with various types of robotic spacecraft, or robotic occupants, which are already either in use or under development for the purpose of exploring remote areas within our own solar system where it is just not feasible to send manned spacecraft.

Before any type of meaningful space exploration can begin, however, the group of physicists would be wise to develop an accurate system of navigation, because our universe (as we see it) is a very large place. There are perhaps billions of galaxies and each galaxy contains billions of stars. One miscalculation, and the spacecraft, in all probability, would become hopelessly lost, so accurate and reliable navigation for this type of space travel must be developed in great detail.

In order to develop an accurate navigation system for this type of space travel, we will first need to develop a thorough understanding of our four-dimensional space-time energy field, namely, the Permittivity and Permeability of free space, so we will have a better understanding of what it is that we are about to do, and how to go about it.

In the first chapter, we discussed the "Big Bang Theory", and how, at the first instant, the primordial singularity first began expanding outward. According to this theory, the primordial singularity started out very small (see page 20) and then expanded outward in all directions (roughly spherical) at an extremely high velocity.

In Chapter 2, we saw how this process is related to the "time" dimension and the three "spatial" dimensions of four-dimensional space-time (see page 42). As you can see on the "Time Chart", with greater and greater amounts of space-time distortion, you tend to move further and further back along the "time line" toward the point of "basic reality". Actually, the whole "time line" for your reference frame is "compressed" (from "where we are now" all the way back to the primordial singularity) in direct proportion to the amount of space-time distortion which has been created.

At the point of "basic reality", the size of our entire universe is contracted down to a singularity with extremely small dimensions, 1.65×10^{-13} centimeter (as we see it), and time is completely stopped.

When the electromagnetic waves within the "Resonator" are brought to resonance, powerful standing waves will begin to develop, and our space-time continuum inside the spacecraft will become increasingly distorted. At the nodes of maximum [positive electric charge/north magnetic pole], and maximum [negative electric charge/south magnetic pole], the Permittivity and Permeability values will soon become saturated and reach their maximum limits for that area of space.

The Permittivity of free space is described as: "The value of proportionality between electric flux intensity and electric field intensity" (capacitance):

ε_0 = 8.854,187,817,6E-12 Farads per meter

The Permeability of free space is described as: "The value of proportionality between magnetic flux density and magnetic field intensity" (inductance) (the $^{-1}$ exponent indicates that there is an inverse relationship between the Permittivity and Permeability values):

μ_0 = 1.256,637,061,44E-7 Henry's per meter^{-1} (The $^{-1}$ exponent means that ε_0 and μ_0 are inversely related)

Any person or material object within the "Resonator" would experience an increasingly sharp reduction in the "effects" of our four-dimensional space-time energy field, and the atoms and molecules of which that person, or object, is composed of would eventually vanish from this reality (time and place). The "time lock" on the atoms and molecules (which are all composed of neutrons and their component parts) would be released and they would be transported almost instantly to another location and then "reappear" based upon the configuration and the orientation of the electric and magnetic fields (standing waves) within the resonator.

Now, let us take another look at the "Space" and "Time" Charts. As you can see on the "Time Chart", whenever you create a certain amount of four-dimensional space-time distortion, the objects or "subjects" within the standing waves inside the spacecraft would tend to move back along the "time line" toward the point of "basic reality". How far they move back along the "time line" depends upon the relative strength of the standing waves, and/or their proximity to and location within the standing wave(s) inside the spacecraft.

This gives you the "time" characteristics of the area (inside the spacecraft) of space-time distortion that you have just created. To calculate the "space" characteristics inside the spacecraft, you take the amount of "time" distortion that you calculated previously, and figure the size of the universe as it was back then, using the "space" chart. Then you superimpose the "Space Chart" over the "Time Chart" with the center of the "Space Chart" directly over the point of "basic reality". Once this relationship is established, you can then determine exactly what types of resonant electromagnetic wave distortions and orientations will be required to get the spacecraft including the "occupants" exactly where you want them to go.

As the size of our universe continued to increase as it expanded, the space within our universe expanded or "stretched" accordingly. In chapter 2, we saw how the images in the mirror, and consequently, how the space within our universe is expanding proportionately as the outer reaches of our universe continue to recede (expand).

This effect will also continue to influence the relative strength of the standing wave(s) inside the spacecraft as it is reduced, and the Permittivity and Permeability values begin to return to normal. Consequently, the space craft, and the "occupants" within it, will return to the present time again, but maybe billions of miles away from the Earth.

Creating space-time distortion, and then using it to travel to distant parts of our universe has become known to physicists as travelling through a "cosmic wormhole". This term was first coined by Professor John Archibald Wheeler (1911 – 2008) at the Institute of Theoretical Physics in Austin, Texas. The advantages of using a spacecraft equipped with space-time distortion technology, as opposed to conventional spacecraft, should be obvious by now. However, we are only just getting started!

In Chapter 3, we determined the ways in which acceleration, velocity, and gravity are related. Acceleration produces velocity-related space-time distortion (relative motion), and consequently, velocity-related space-time distortion produces acceleration (gravity). In this chapter, we have seen how an electro-mechanical device (resonator) can produce various amounts of velocity-related space-time distortion.

Now, if our group of physicists would be able to redesign their spacecraft so that the resonant electromagnetic waves which produced the space-time distortion inside their spacecraft, could be transmitted across the outer surface of the hull, and controlled so it would produce varying degrees of space-time distortion around the outside of the spacecraft, the physicists would then be able to create space-time distortion on the upper side of their spacecraft in order to produce lift in a gravitational field. They could also produce positive space-time distortion on one side of their spacecraft and negative space-time distortion on the opposite side of their spacecraft in order to accelerate their spacecraft, and, consequently, produce relative motion in a certain direction.

In addition to that, by utilizing the space-time distortion principles set forth in this chapter, and then by utilizing the "Space Chart" in conjunction with the "Time Chart" as a means of navigation, it could be possible to travel to almost any point within our universe, and at almost any time you wanted to arrive, past, present, or future.

In chapter 3, it was briefly mentioned that the "real" "pressure-related" expansion of our universe, as it exists within the realm of "basic reality", is actually perpendicular to the "velocity-related" expansion process which is producing our perceptions of our universe (relativity) as we see it. Also, due to the nature of the laws of physics in the "empty black void" (realm of "basic reality"), the "real", "pressure-related" expansion process cannot occur in an even and continuous manner within the realm of "basic reality". It actually occurs in "steps", with each "step", or "Dimension", having its own separate existence within the realm of "basic reality" and representing a different size of the singularity and "pressure" of the physical substance within the singularity (see page 136).

Each of the first fifteen to twenty of these now-separate singularities has produced its own separate universe ("Dimension") within the realm of "basic reality". The total number of these separate "Dimensions", or "universes", can be determined by measuring the difference in "observed mass" and "gravitational mass" in regard to massive objects such as nearby galaxies which can be readily observed and their "gravitational mass" can be calculated fairly accurately. These separate universes are what contains all the "Dark Matter" that we cannot see, but which also has a measurable, cumulative effect on all our various gravitational fields (velocity-related space-time distortion) throughout our universe.

As the "physical substance" (singularity) continued to expand, producing a rather large number of these separate existences, its size soon reached the expansion threshold where the neutron (full-cycle) waveform of which its outer surface consisted, was not highly concentrated enough to produce functional atoms. Even though no further atoms or mass were able to be produced, this expansion process continues, even to this day, and is causing the "passage of time", as we perceive it.

In addition to that, the continuing expansion process contributes to the total strength of velocity-related space-time distortion, which, consequently, is represented by the Permittivity and Permeability values. In Chapter 2, it was mentioned that the continuing expansion process is what produces our perceptions of the passage of time, and is what keeps the stars burning in order to produce a continual source of energy, it is also what really keeps everything in our universe going.

The real physical substance, as it exists in each separate existence ("Dimension") within the realm of "basic reality", is the primordial singularity of each of those particular "Dimensions". As we have observed in our own "Dimension" (our universe as we see it), it is the "pressure" of the physical substance (within the primordial singularity) which determines the speed of light, Planck's Constant, the gravitational constant, the Permittivity and Permeability values, and our perceptions of time, distance, energy, relative motion, and acceleration. Each one of these physical constants and variables have a different value in each of the different "Dimensions", or universes (see page 136), because the size and pressure of the physical substance in the primordial singularity is different in each one of them.

Even though it is essentially the same substance in all of the different "Dimensions", the different sizes and pressure concentrations of each singularity are what gives each one its own separate identity. Surprisingly, we actually have accumulated observational evidence which can enable us to determine approximately how many of these separate singularities (existences) that were able to produce functional atoms which have mass, and therefore, have produced separate universes.

According to some of the most recent astronomical observations, there should be approximately 10 - 15 of these separate existences ("Dimensions") which contain the "dark matter" that we cannot observe directly but they can still be detected because of the overall "effects" that they have on our gravitational fields (velocity-related space-time distortion).

In "Dimensions" (universes) which came into existence before ours, the inhabitants would perceive the speed of light to be greater, the electromagnetic wave/particles would appear smaller, and their whole universe would appear larger than ours. In universes which came into existence after ours, their speed of light would appear to be slower, their atoms would appear larger, and their universe would appear to be smaller than ours.

These different universes remain separate from each other simply because energy cannot travel from one to the other. We are not aware of their existence because their electromagnetic (light) waves are not compatible with the atoms in our universe. In other words, the "fixed" effects of electricity and magnetism are also not quite the same in any of these separate universes.

There are also similarities in each of these separate universes, as well. Since it is the same physical substance of which each separate singularity consists, then the neutron "outer shock-wave" and the subsequent distribution of matter within each primordial singularity will be identical. The distribution of stars, galaxies, and clusters of galaxies should be nearly identical in each of the separate universes and contribute to the overall strength of the four-dimensional space-time energy field which is represented by the Permittivity and Permeability values.

There is also an enormous amount of electromagnetic wave energy which has been, and which is still being produced by all of the stars in all of the galaxies. Most of this energy is still traversing our universe, and the other universes, ever since they came into existence. Our universal four-dimensional space-time energy field is evidently not only conducting (containing) all of this energy just within our own universe, but also within all of the other separate universes ("Dimensions") as well. This is why there appears to be so much "Dark Energy" within our universe.

The "Time" and "Space Charts" are able to describe the "time" and "space" characteristics of each different universe ("Dimension"). The only things that are actually different within each different universe are the "size" and "pressure" of the physical substance (primordial singularity) as it exists in each different "Dimension" (universe) within the realm of "basic reality".

Also, the length of the "time line" and the "space" characteristics (size) are different within each "Dimension" which causes each "Dimension" to "appear" to be expanding at a different rate, and time would also proceed at a slightly different rate according to observers in each different "Dimension".

If our group of physicists could be able to control the frequencies of the resonant electromagnetic waves on the outer hull of their spacecraft in order to provide the proper amount of "pressure-related" space-time distortion, in addition to the electromagnetic wave resonance inside the spacecraft, this new type of spacecraft would now be equipped for inter-dimensional travel as well as time and space travel.

If this new technology can be perfected, can you just imagine the endless number of possibilities that this field of scientific research would open up for us, and what this would mean for the future of space exploration.

The "Time" and "Space Charts" are the keys to both a new type of space travel, and a better understanding of our universe from the time it came into existence approximately 15 billion years ago up until the present day. They describe what the component parts of atoms really are, and how energy is transferred from one place to another within our universe. And most importantly, the time and space charts provide us with a link between the Theory of Relativity and Quantum Theory.

As the pieces of this great puzzle continue to fall into place, we can now see that the future holds many more interesting discoveries in store for us.

CHAPTER 7

THE FUTURE OF OUR UNIVERSE

Now that we have a better understanding of what has happened to our universe in the past, and what is going on in the present, let's try and speculate about what may happen to our universe in the future. We can start by attempting to develop a better understanding of the details regarding the internal structure of the singularity and how it has produced our four-dimensional space-time continuum (the reality that we perceive).

As we have seen in the previous chapters, all the forces of nature, the atoms, and all quantum processes are all intimately linked and related to each other through the four-dimensional space-time energy field which is represented by the Permittivity and Permeability of free space through the singularity (see page 169). The "Big Bang Theory" would indicate that our universe (four-dimensional space-time) began with the primordial singularity, and has been expanding outward at an extremely high velocity ever since.

According to this theory, time, distance, relative velocity, mass, and acceleration (gravity), can all be compared to a huge, three-dimensional, holographic image which is being produced by the effects of our four-dimensional space-time energy field which is represented by the Permittivity and Permeability values. Also, these concepts are all meaningless in the realm of "basic reality" -- the empty black void in which our universe exists. This void can be considered to be truly empty space, or "dead space". From the standpoint of "basic reality" (the void), the expansion of our universe has never "really" taken place.

A closer examination of the physical properties and internal structure of the primordial singularity, which is represented by several key physical and numerical constants, may help us to better-understand the relationships between atoms and our universe as a whole. A numerical analysis of several of these important physical constants has revealed what appears to be a well-defined internal structure within the singularity which also closely resembles many of the internal characteristics of atoms as well.

(Continued on page 170)

NUMERICAL REPRESENTATION OF A SPHERE

The sphere is the most fundamental geometric form which represents our universe, and many of the things within it on the macro scale and also on the micro scale. This demonstrates that **4π** is the mathematical common denominator upon which our universe and everything within it is unified including gravity, electricity, magnetism, energy, wave motion, and atomic relationships according to the following examples:

Since $c = \dfrac{\alpha \hbar}{e^2} = 299{,}792{,}458$ m/sec $= \dfrac{1}{\sqrt{\varepsilon_0 \mu_0}}$; and $\dfrac{e^2}{\alpha \hbar} = \sqrt{\varepsilon_0 \mu_0}$; then $\dfrac{e^4}{\alpha^2 \hbar^2 \varepsilon_0} = $ **4π**

Since $\dfrac{1}{c} = \dfrac{e^2}{\alpha \hbar} = 3.335{,}640{,}951{,}981{,}52\text{E-}9$ m/sec $= \sqrt{\varepsilon_0 \mu_0}$; then $\dfrac{e^4}{\alpha^2 \hbar^2 \varepsilon_0} = $ **4π**

Since $c^2 = \dfrac{\alpha^2 \hbar^2}{e^4} = 89{,}875{,}517{,}873{,}681{,}764$ m/sec $= \dfrac{1}{\varepsilon_0 \mu_0}$; then $\dfrac{e^4}{\alpha^2 \hbar^2 \varepsilon_0} = $ **4π**

Since $\dfrac{1}{c^2} = \dfrac{e^4}{\alpha^2 \hbar^2} = 1.112{,}650{,}056{,}053{,}62\text{E-}17$ m/sec $= \varepsilon_0 \mu_0$; then $\dfrac{e^4}{\alpha^2 \hbar^2 \varepsilon_0} = $ **4π**

Since $\varepsilon_0 = \dfrac{e^4}{\alpha^2 \hbar^2 \mu_0} = 8.854{,}187{,}817{,}62\text{E-}12$ Farads/meter $= \dfrac{1}{c^2 \mu_0}$; then $\dfrac{e^4}{\alpha^2 \hbar^2 \varepsilon_0} = $ **4π**

Since $\mu_0 = \dfrac{e^4}{\alpha^2 \hbar^2 \varepsilon_0} = 1.256{,}637{,}061{,}44\text{E-}7$ Henry's/meter $= \dfrac{1}{c^2 \varepsilon_0}$; then $\dfrac{e^4}{\alpha^2 \hbar^2 \varepsilon_0} = $ **4π**

Energy is related to harmonic (wave) motion which is also represented by: **4π**

*ESU - Electrostatic System of Units (see Table of Constants on page 174 for legend)

ESU of Potential = 299,792,458/1.000,000 = 299.792,458 Volts
ESU of Resistance = 299,792,458²/100,000 = 898,755,178,736.817,64 Ohms
ESU of Current = 1/299,792,458/10 = 3.335,640,951,981,52E-10 Amperes
ESU of Conductance = (1/299,792,458²) X 100,000 = 1.112,650,056,053,62E-12 Siemens
ESU of Capacitance = 1/299,792,458² X 100,000 = 1.112,650,056,053,62E-12 Farads
ESU of Inductance = 299,792,458²/100,000 = 898,755,178,736.817,64 Henrys
ESU of Quantity = 1/299,792,458/10 = 3.335,640,951,981,52E-10 Coulombs

-----

Volt = 1/(299,792,458/1,000,000) = 3.335,640,951,981,52E-03 ESU of Potential
Ohm = 1/(299,792,458²/100,000) = 1.112,650,056,053,62E-12 ESU of Resistance
Amperes = 1/(1/299,792,458)/10 = 2,997,924,580 ESU of Current
Siemens = 1(1/299,792,458² X 100,000 = 898,755,178,736.817,64 ESU of Conductance
Farads = 1/(1/299,792,458² X 100,000) = 898,755,178,736.817,64 ESU of Capacitance
Henrys = 1/(299,792,458²/100,000) = 1.112,650,056,053,62E-12 ESU of Inductance
Coulombs = 1/(1/299,792,458)/10 = 2,997,924,580 ESU of Quantity

(Continued from page 168)

By comparing the information on pages 169 and 176, it becomes obvious that there is a comprehensive, underlying set of principles upon which the fundamental forces within our universe are based. A closer analysis of the three main physical constants: the elementary charge squared (e^2); the reduced Planck constant ($\hbar$); and the fine structure constant (α), reveals that, taken together, these values form the basic electromagnetic wave structure upon which everything (mass) within our universe is based, both at the atomic level and at the astronomical level.

The speed of light (c), which represents the actual physical properties of the Permittivity and Permeability values, determines the strength of electric, magnetic and gravitational fields, conducts the electromagnetic waves through space, and facilitates the various atomic and subatomic interactions. This is why (c) is so prominent in calculations regarding physical and atomic quantities. It is also interesting to note that while e^2, α, and $\hbar$ are always constant, or held in a "fixed state", the Permittivity and Permeability values appear to play both roles -- they act as constants in many cases, but they can also serve as variables under certain circumstances regarding relative motion, electric, magnetic, and gravitational fields.

Regarding the "fixed" constants, the elementary charge (e) appears to have its own unique mathematical orientation and it is always squared in comparison to the other two constants (α) and ($\hbar$). This is most likely due to the fact that there are equal-but-opposite positive and negative distortions in the physical substance which the electric charge represents, both of which are equal to (e).

The fine structure constant (α) represents the strength of coupling between the basic elementary charges at the atomic level, and (ℏ) is the angular momentum which represents the rolling or twisting motion that the basic electromagnetic wave would have if it were not being held in a "fixed state" by the laws of physics in the "void" (or realm of "basic reality").

Another significant physical process which originates within the singularity is the inverse–square relationship which is produced by the four different physical orientations, $\frac{\alpha \hbar}{e^2}$, $\frac{e^2}{\alpha \hbar}$, $\frac{\alpha^2 \hbar^2}{e^4}$, $\frac{e^4}{\alpha^2 \hbar^2}$, of the three primary constants which are then represented into our physical reality by the Permittivity and Permeability values (see page 78). This is the basic underlying physical process which produces the effects of relativity (relativity equations) and the 1/r² Law (see pages 15, and 71).

In addition to producing the effects of relativity, the basic structural orientation of the three primary constants within the primordial singularity, $\frac{e^2}{\alpha \hbar}$, also provides the physical basis for Ohms Law in regard to electricity and magnetism.

From the information on pages 169, 174, 177, 178, 179 and 180, you can see that the basic structural orientation of these primary physical constants, which is then represented into our reality by the physical properties of the Permittivity and Permeability values, can then be mathematically reduced to 4π, which is the numerical representation of the spherical primordial singularity. This is a very interesting phenomenon from both a mathematical and a physical standpoint. What we are seeing here is the underlying physical process by which all electromagnetic field relationships, all the space-time (relativistic) phenomena, and all quantum physical processes within our universe are based.

The information on the pages listed above also represents much of the mathematical evidence which is necessary to prove this theory. A thorough and comprehensive analysis of this data should better-enable us to fully interpret the results of quantum experiments, improve many areas of science and technology, improve our means of space travel, and also help us determine what may happen to our universe in the future.

In Chapter 5, we discussed quantum experiments in which the "sphere of influence" (wave function) of a photon, or a neutron, could be made to collapse merely by the act of observing it. Well, our universe is the "sphere of influence" of the primordial singularity, and it is only a question of how long it will take for the physical substance, which is driving (causing) the expansion of our universe, to reach its threshold of maximum expansion.

As the physical substance approaches its threshold of maximum expansion, (see page 15) the inability of the physical substance to continue becoming thinner and less concentrated will probably have some "braking effect" on the overall expansion process which should cause a significant slow-down just prior to coming to a complete stop. At that point, everything would probably just "freeze" up.

There would be no more passage of time, relative motion, energy, or acceleration (gravity). Without the passage of time, the stars would all stop burning and collapse. The Permittivity and Permeability values would suddenly cease to exist and would no longer be able to conduct the electromagnetic waves in any direction, and so the "lights" would probably go out" rather quickly.

Of course, at the present time, this is all just a matter of conjecture. But by the same token, it may be possible that the physical expansion process of our universe could continue for billions of years into the future. According to our latest observations, the expansion process which our universe is currently undergoing, still appears to be quite robust and some astronomers are even claiming that it may be speeding up.

However, one of the main reasons that a quantum (wave function) collapse of our universe may occur eventually is that the four-dimensional space-time energy field, which represents the combined wave functions of all the neutrons, including electrons and protons (mass), and which is represented by the Permittivity and Permeability values, is part of an indivisible whole.

(Continued on page 181)

TABLE OF CONSTANTS

© 2015 Charles R. Storey all rights reserved
*For greater resolution see "High Precision Table of Constants" in the Appendix.

Elementary charge [4] $e^4 = \alpha^2 \hbar^2 \varepsilon_0 \mu_0$ = 6.589,333,663,690,559,907,896,212,880,112E-90 C

Elementary charge² $e^2 = \dfrac{\alpha \hbar}{c} = \alpha \hbar \sqrt{\varepsilon_0 \mu_0}$ = 2.566,969,743,431,067,383,000,001,685,533,2E-38 C

Reduced Planck cnst $\hbar = \dfrac{c e^2}{\alpha} = \dfrac{e^2}{\alpha \sqrt{\varepsilon_0 \mu_0}}$ = 1.054,571,725,339,976,234,000,102,692,452,8E-34 J·S

Fine structure constant $\alpha = \dfrac{c e^2}{\hbar} = \dfrac{e^2}{\hbar \sqrt{\varepsilon_0 \mu_0}}$ = .007,297,352,569,800,184,997,640,004,791,553,6

Speed of light $c = \dfrac{\alpha \hbar}{e^2} = \dfrac{1}{\sqrt{\varepsilon_0 \mu_0}}$ = 299,792,458 meters/second

1/the speed of light $\dfrac{1}{c} = \dfrac{e^2}{\alpha \hbar} = \sqrt{\varepsilon_0 \mu_0}$ = 3.335,640,951,981,520,495,755,767,144,749,2E-9 m/sec

Speed of light squared $c^2 = \dfrac{\alpha^2 \hbar^2}{e^4} = \dfrac{1}{\varepsilon_0 \mu_0}$ = 89,875,517,873,681,764 meters/second

1/the speed of light squared $\dfrac{1}{c^2} = \dfrac{e^4}{\alpha^2 \hbar^2} = \varepsilon_0 \mu_0$ = 1.112,650,056,053,618,432,174,089E-17 m/sec

$\varepsilon_0 = \dfrac{1}{c^2 \mu_0} = \dfrac{e^4}{\alpha^2 \hbar^2 \mu_0}$ = 8.854,187,817,620,389,850,536,563,031,710,9E-12 Farads/meter

$\mu_0 = \dfrac{1}{c^2 \varepsilon_0} = \dfrac{e^4}{\alpha^2 \hbar^2 \varepsilon_0}$ = 12.566,370,614,359,172,953,850,573,533,118E-7 Henry's/meter

*The Permeability of free space can be numerically represented by 4π (12.566,370,614,359).

Within the primordial singularity very small numbers can be numerically equal to very large numbers because the expansion of our universe which has produced our four dimensional space-time continuum (the standards by which we measure time and distance) has not taken place yet. Within the realm in which this singularity exists, the realm of "basic reality", the dimensions of space (distance) and time, as we perceive them, are meaningless.

These numerical equalities can be used to describe the basic structure and respective orientations of the physical properties within the singularity that set the standards by which the properties of electricity and magnetism, electromagnetic waves, electromagnetic wave/particles of which atoms consist, and our entire four-dimensional space-time continuum are patterned:

$$\alpha \hbar = \frac{e^2}{\sqrt{\varepsilon_0 \mu_0}}$$

$$\alpha^2 \hbar^2 = \frac{e^4}{\varepsilon_0 \mu_0}$$

$$\frac{e^2}{\alpha} = \hbar\sqrt{\varepsilon_0 \mu_0}$$

$$\frac{e^4}{\alpha^2} = \hbar^2 \varepsilon_0 \mu_0$$

$$\frac{e^2}{\hbar} = \alpha\sqrt{\varepsilon_0 \mu_0}$$

$$\frac{e^4}{\hbar^2} = \alpha^2 \varepsilon_0 \mu_0$$

$$\frac{\alpha \hbar}{\sqrt{\varepsilon_0 \mu_0}} = \frac{e^2}{\varepsilon_0 \mu_0}$$

$$\frac{\alpha}{\sqrt{\varepsilon_0 \mu_0}} = \frac{e^2}{\hbar \varepsilon_0 \mu_0}$$

$$\frac{\hbar}{\sqrt{\varepsilon_0 \mu_0}} = \frac{e^2}{\alpha \varepsilon_0 \mu_0}$$

Numerically equal expressions of the four primary constants:

$$\frac{e^2}{\hbar} = \alpha\sqrt{\varepsilon_0 \mu_0} = \frac{\alpha}{c}$$

$$\frac{e^2}{\alpha} = \hbar\sqrt{\varepsilon_0 \mu_0} = \frac{\hbar}{c}$$

$$\frac{e^4}{\hbar^2} = \alpha^2 \varepsilon_0 \mu_0 = \frac{\alpha^2}{c^2}$$

$$\frac{e^4}{\alpha^2} = \hbar^2 \varepsilon_0 \mu_0 = \frac{\hbar^2}{c^2}$$

Cross-multiplication produces two numerically equal expressions from the examples above, resulting in the final equations:

$$\alpha \hbar = \frac{e^2}{\sqrt{\varepsilon_0 \mu_0}} = c e^2$$

and

$$\alpha^2 \hbar^2 = \frac{e^4}{\varepsilon_0 \mu_0} = c^2 e^4$$

thus

$$\frac{\alpha \hbar}{c e^2} = 1 = \frac{c e^2}{\alpha \hbar}$$

and

$$\frac{\alpha^2 \hbar^2}{c^2 e^4} = 1 = \frac{c^2 e^4}{\alpha^2 \hbar^2}$$

*A more detailed description of the steps in the equations is provided on the following page.

Since $\quad \alpha \hbar \;=\; c e^2 \;=\; \dfrac{e^2}{\sqrt{\varepsilon_0 \mu_0}}\;$; then $\quad \dfrac{e^4}{\alpha^2 \hbar^2 \varepsilon_0} \;=\; 4\pi$

Since $\quad \dfrac{e^2}{\hbar} \;=\; \dfrac{\alpha}{c} \;=\; \alpha\sqrt{\varepsilon_0 \mu_0}\;$; then $\quad \dfrac{e^4}{\alpha^2 \hbar^2 \varepsilon_0} \;=\; 4\pi$

Since $\quad \dfrac{e^2}{\alpha} \;=\; \dfrac{\hbar}{c} \;=\; \hbar\sqrt{\varepsilon_0 \mu_0}\;$; then $\quad \dfrac{e^4}{\alpha^2 \hbar^2 \varepsilon_0} \;=\; 4\pi$

Since $\quad \dfrac{1}{\alpha \hbar} \;=\; \dfrac{1}{c e^2} \;=\; \dfrac{\sqrt{\varepsilon_0 \mu_0}}{e^2}\;$; then $\quad \dfrac{e^4}{\alpha^2 \hbar^2 \varepsilon_0} \;=\; 4\pi$

Since $\quad \dfrac{\hbar}{e^2} \;=\; \dfrac{c}{\alpha} \;=\; \dfrac{1}{\alpha\sqrt{\varepsilon_0 \mu_0}}\;$; then $\quad \dfrac{e^4}{\alpha^2 \hbar^2 \varepsilon_0} \;=\; 4\pi$

Since $\quad \dfrac{\alpha}{e^2} \;=\; \dfrac{c}{\hbar} \;=\; \dfrac{1}{\hbar\sqrt{\varepsilon_0 \mu_0}}\;$; then $\quad \dfrac{e^4}{\alpha^2 \hbar^2 \varepsilon_0} \;=\; 4\pi$

← entropy toward unity →

Since $\quad \alpha^2 \hbar^2 \;=\; c^2 e^4 \;=\; \dfrac{e^4}{\varepsilon_0 \mu_0}\;$; then $\quad \dfrac{e^4}{\alpha^2 \hbar^2 \varepsilon_0} \;=\; 4\pi$

Since $\quad \dfrac{e^4}{\hbar^2} \;=\; \dfrac{\alpha^2}{c^2} \;=\; \alpha^2 \varepsilon_0 \mu_0\;$; then $\quad \dfrac{e^4}{\alpha^2 \hbar^2 \varepsilon_0} \;=\; 4\pi$

Since $\quad \dfrac{e^4}{\alpha^2} \;=\; \dfrac{\hbar^2}{c^2} \;=\; \hbar^2 \varepsilon_0 \mu_0\;$; then $\quad \dfrac{e^4}{\alpha^2 \hbar^2 \varepsilon_0} \;=\; 4\pi$

Since $\quad \dfrac{1}{\alpha^2 \hbar^2} \;=\; \dfrac{1}{c^2 e^4} \;=\; \dfrac{\varepsilon_0 \mu_0}{e^4}\;$; then $\quad \dfrac{e^4}{\alpha^2 \hbar^2 \varepsilon_0} \;=\; 4\pi$

Since $\quad \dfrac{\hbar^2}{e^4} \;=\; \dfrac{c^2}{\alpha^2} \;=\; \dfrac{1}{\alpha^2 \varepsilon_0 \mu_0}\;$; then $\quad \dfrac{e^4}{\alpha^2 \hbar^2 \varepsilon_0} \;=\; 4\pi$

Since $\quad \dfrac{\alpha^2}{e^4} \;=\; \dfrac{c^2}{\hbar^2} \;=\; \dfrac{1}{\hbar^2 \varepsilon_0 \mu_0}\;$; then $\quad \dfrac{e^4}{\alpha^2 \hbar^2 \varepsilon_0} \;=\; 4\pi$

4π is the spherical representation of the Permeability of free space **μ₀**

$$\alpha \hbar = \frac{e^2}{\sqrt{\varepsilon_0 \mu_0}} \;;\; \frac{\alpha \hbar}{e^2} = \frac{1}{\sqrt{\varepsilon_0 \mu_0}} \;;\; \frac{e^2}{\alpha \hbar} = \sqrt{\varepsilon_0 \mu_0} \;;\; \frac{e^4}{\alpha^2 \hbar^2} = \varepsilon_0 \mu_0 \;;\; \frac{e^4}{\alpha^2 \hbar^2 \varepsilon_0} = \mu_0$$

$$\frac{e^2}{\hbar} = \alpha \sqrt{\varepsilon_0 \mu_0} \;;\; \quad\quad\quad \frac{e^2}{\alpha \hbar} = \sqrt{\varepsilon_0 \mu_0} \;;\; \frac{e^4}{\alpha^2 \hbar^2} = \varepsilon_0 \mu_0 \;;\; \frac{e^4}{\alpha^2 \hbar^2 \varepsilon_0} = \mu_0$$

$$\frac{e^2}{\alpha} = \hbar \sqrt{\varepsilon_0 \mu_0} \;;\; \quad\quad\quad \frac{e^2}{\alpha \hbar} = \sqrt{\varepsilon_0 \mu_0} \;;\; \frac{e^4}{\alpha^2 \hbar^2} = \varepsilon_0 \mu_0 \;;\; \frac{e^4}{\alpha^2 \hbar^2 \varepsilon_0} = \mu_0$$

$$\frac{1}{\alpha \hbar} = \frac{\sqrt{\varepsilon_0 \mu_0}}{e^2} \;;\; \quad\quad\quad \frac{e^2}{\alpha \hbar} = \sqrt{\varepsilon_0 \mu_0} \;;\; \frac{e^4}{\alpha^2 \hbar^2} = \varepsilon_0 \mu_0 \;;\; \frac{e^4}{\alpha^2 \hbar^2 \varepsilon_0} = \mu_0$$

$$\frac{\hbar}{e^2} = \frac{1}{\alpha \sqrt{\varepsilon_0 \mu_0}} \;;\; \frac{\alpha \hbar}{e^2} = \frac{1}{\sqrt{\varepsilon_0 \mu_0}} \;;\; \frac{e^2}{\alpha \hbar} = \sqrt{\varepsilon_0 \mu_0} \;;\; \frac{e^4}{\alpha^2 \hbar^2} = \varepsilon_0 \mu_0 \;;\; \frac{e^4}{\alpha^2 \hbar^2 \varepsilon_0} = \mu_0$$

$$\frac{\alpha}{e^2} = \frac{1}{\hbar \sqrt{\varepsilon_0 \mu_0}} \;;\; \frac{\alpha \hbar}{e^2} = \frac{1}{\sqrt{\varepsilon_0 \mu_0}} \;;\; \frac{e^2}{\alpha \hbar} = \sqrt{\varepsilon_0 \mu_0} \;;\; \frac{e^4}{\alpha^2 \hbar^2} = \varepsilon_0 \mu_0 \;;\; \frac{e^4}{\alpha^2 \hbar^2 \varepsilon_0} = \mu_0$$

← entropy   direction in which equations are solved (toward unity) →

$$\alpha^2 \hbar^2 = \frac{e^4}{\varepsilon_0 \mu_0} \;;\; \frac{\alpha^2 \hbar^2}{e^4} = \frac{1}{\varepsilon_0 \mu_0} \;;\; \quad\quad\quad \frac{e^4}{\alpha^2 \hbar^2} = \varepsilon_0 \mu_0 \;;\; \frac{e^4}{\alpha^2 \hbar^2 \varepsilon_0} = \mu_0$$

$$\frac{e^4}{\hbar^2} = \alpha^2 \varepsilon_0 \mu_0 \;;\; \quad\quad\quad\quad\quad \frac{e^4}{\alpha^2 \hbar^2} = \varepsilon_0 \mu_0 \;;\; \frac{e^4}{\alpha^2 \hbar^2 \varepsilon_0} = \mu_0$$

$$\frac{e^4}{\alpha^2} = \hbar^2 \varepsilon_0 \mu_0 \;;\; \quad\quad\quad\quad\quad \frac{e^4}{\alpha^2 \hbar^2} = \varepsilon_0 \mu_0 \;;\; \frac{e^4}{\alpha^2 \hbar^2 \varepsilon_0} = \mu_0$$

$$\frac{1}{\alpha^2 \hbar^2} = \frac{\varepsilon_0 \mu_0}{e^4} \;;\; \quad\quad\quad\quad\quad \frac{e^4}{\alpha^2 \hbar^2} = \varepsilon_0 \mu_0 \;;\; \frac{e^4}{\alpha^2 \hbar^2 \varepsilon_0} = \mu_0$$

$$\frac{\hbar^2}{e^4} = \frac{1}{\alpha^2 \varepsilon_0 \mu_0} \;;\; \frac{\alpha^2 \hbar^2}{e^4} = \frac{1}{\varepsilon_0 \mu_0} \;;\; \quad\quad\quad \frac{e^4}{\alpha^2 \hbar^2} = \varepsilon_0 \mu_0 \;;\; \frac{e^4}{\alpha^2 \hbar^2 \varepsilon_0} = \mu_0$$

$$\frac{\alpha^2}{e^4} = \frac{1}{\hbar^2 \varepsilon_0 \mu_0} \;;\; \frac{\alpha^2 \hbar^2}{e^4} = \frac{1}{\varepsilon_0 \mu_0} \;;\; \quad\quad\quad \frac{e^4}{\alpha^2 \hbar^2} = \varepsilon_0 \mu_0 \;;\; \frac{e^4}{\alpha^2 \hbar^2 \varepsilon_0} = \mu_0$$

*Numerical orientations of the physical properties within the primordial singularity.

$$\alpha \hbar = c e^2 = \frac{e^2}{\sqrt{\varepsilon_0 \mu_0}} = 7.695,581,656 = \frac{1}{1.299,446,935} = \sqrt{5.9221,976,97} = \frac{1}{\sqrt{1.688,562,339}}$$

$$\frac{1}{\alpha \hbar} = \frac{1}{c e^2} = \frac{\sqrt{\varepsilon_0 \mu_0}}{e^2} = 1.299,446,935 = \frac{1}{7.695,581,656} = \sqrt{1.688,562,339} = \frac{1}{\sqrt{59.221,976,97}}$$

$$\alpha^2 \hbar^2 = c^2 e^2 = \frac{e^4}{\varepsilon_0 \mu_0} = 59.221,976,97 = \frac{1}{1.688,562,339} = 7.695,581,656^2 = \frac{1}{1.299,446,935^2}$$

$$\frac{1}{\alpha^2 \hbar^2} = \frac{1}{c^2 e^4} = \frac{\varepsilon_0 \mu_0}{e^4} = 1.688,562,339 = \frac{1}{5.922,197,697} = 1.299,446,935^2 = \frac{1}{7.695,581,656^2}$$

$$\frac{e^2}{\hbar} = \frac{\alpha}{c} = \alpha\sqrt{\varepsilon_0 \mu_0} = 2.434,134 = \frac{1}{4,108.235,917,7} = \sqrt{5.925,012,22} = \frac{1}{\sqrt{16,877,602.355}}$$

$$\frac{\hbar}{e^2} = \frac{c}{\alpha} = \frac{1}{\alpha\sqrt{\varepsilon_0 \mu_0}} = 4,108.235,917,7 = \frac{1}{2.434,134} = \sqrt{1.687,760,235} = \frac{1}{\sqrt{5.925,012,207}}$$

$$\frac{e^4}{\hbar^2} = \frac{\alpha^2}{c^2} = \alpha^2 \varepsilon_0 \mu_0 = 5.925,012,3 = \frac{1}{1.687,760,23} = 2.434,134,796^2 = \frac{1}{41,082,359,177^2}$$

$$\frac{\hbar^2}{e^4} = \frac{c^2}{\alpha^2} = \frac{1}{\alpha^2 \varepsilon_0 \mu_0} = 1.687,760,23 = \frac{1}{5.925,012,207} = 41,082,359,177^2 = \frac{1}{2.434,134,796^2}$$

$$\frac{e^2}{\alpha} = \frac{\hbar}{c} = \hbar \sqrt{\varepsilon_0 \mu_0} = 3.517,672,61 = \frac{1}{2.842,788,69} = \sqrt{1.237,402,049} = \frac{1}{\sqrt{8.081,477,6}}$$

$$\frac{\alpha}{e^2} = \frac{c}{\hbar} = \frac{1}{\hbar \sqrt{\varepsilon_0 \mu_0}} = 2.842,788,709 = \frac{1}{3.517,672,61} = \sqrt{8.081,447,774} = \frac{1}{\sqrt{1.237,402,049}}$$

$$\frac{e^4}{\alpha^2} = \frac{\hbar^2}{c^2} = \hbar^2 \varepsilon_0 \mu_0 = 1.237,402,063 = \frac{1}{8.081,447,744} = 3.517,672,61^2 = \frac{1}{2.842,788,709^2}$$

$$\frac{\alpha^2}{e^4} = \frac{c^2}{\hbar^2} = \frac{1}{\hbar^2 \varepsilon_0 \mu_0} = 8.081,447,6 = \frac{1}{1.237,402,049} = 2.842,788,69^2 = \frac{1}{3.517,672,641^2}$$

*Orientations of the electromagnetic properties within the primordial singularity.

The speed of light $\quad c \;=\; \dfrac{\alpha \hbar}{e^2} \;=\; \dfrac{1}{\sqrt{\varepsilon_0 \mu_0}} \;=\; 299{,}792{,}458$ meters per second

1/the speed of light $\quad \dfrac{1}{c} \;=\; \dfrac{e^2}{\alpha \hbar} \;=\; \sqrt{\varepsilon_0 \mu_0} \;=\; 3.335{,}640{,}951{,}981{,}52\text{E-}9$ m/sec

Speed of light squared $\quad c^2 \;=\; \dfrac{\alpha^2 \hbar^2}{e^4} \;=\; \dfrac{1}{\varepsilon_0 \mu_0} \;=\; 89{,}875{,}517{,}873{,}681{,}764$

1/the speed of light squared $\quad \dfrac{1}{c^2} \;=\; \dfrac{e^4}{\alpha^2 \hbar^2} \;=\; \varepsilon_0 \mu_0 \;=\; 1.112{,}650{,}056{,}053{,}62\text{E-}17$ m/sec

Coulomb constant $\quad k_e \;=\; \dfrac{\alpha^2 \hbar^2}{e^4} \;=\; \dfrac{1}{\varepsilon_0 \mu_0} \;=\; 8.987{,}551{,}787{,}368{,}176{,}4\text{E}9$ N·m²C⁻¹

$299{,}792{,}458 \;=\; \dfrac{1}{3.335{,}640{,}951{,}981{,}52} \;=\; \sqrt{89{,}875{,}517{,}873{,}681{,}764} \;=\; \dfrac{1}{\sqrt{1.112{,}650{,}056{,}053{,}62}}$

$3.335{,}640{,}951{,}981{,}52 \;=\; \dfrac{1}{299{,}792{,}458} \;=\; \sqrt{1.112{,}650{,}056{,}053{,}62} \;=\; \dfrac{1}{\sqrt{89{,}875{,}517{,}873{,}681{,}764}}$

$1.112{,}650{,}056{,}053{,}62 \;=\; \dfrac{1}{89{,}875{,}517{,}873{,}681{,}764} \;=\; 3.335{,}640{,}951{,}981{,}52^2 \;=\; \dfrac{1}{\sqrt{299{,}792{,}458}}$

$89{,}875{,}517{,}873{,}681{,}764 \;=\; \dfrac{1}{1.112{,}650{,}056{,}053{,}62} \;=\; 299{,}792{,}458^2 \;=\; \dfrac{1}{\sqrt{3.335{,}640{,}951{,}981{,}52}}$

*ESU - Electrostatic System of Units

ESU of Potential = 299,792,458/1.000,000 = 299.792,458 Volts
ESU of Resistance = 299,792,458²/100,000 = 898,755,178,736.817,64 Ohms
ESU of Current = 1/299,792,458/10 = 3.335,640,951,981,52E-10 Amperes
ESU of Conductance = (1/299,792,458²) X 100,000 = 1.112,650,056,053,62E-12 Siemens
ESU of Capacitance = 1/299,792,458² X 100,000 = 1.112,650,056,053,62E-12 Farads
ESU of Inductance = 299,792,458²/100,000 = 898,755,178,736.817,64 Henrys
ESU of Quantity = 1/299,792,458/10 = 3.335,640,951,981,52E-10 Coulombs

---

Volt = 1/(299,792,458/1,000,000) = 3.335,640,951,981,52E-03 ESU of Potential
Ohm = 1/(299,792,458²/100,000) = 1.112,650,056,053,62E-12 ESU of Resistance
Amperes = 1/(1/299,792,458)/10 = 2,997,924,580 ESU of Current
Siemens = 1(1/299,792,458² X 100,000 = 898,755,178,736.817,64 ESU of Conductance
Farads = 1/(1/299,792,458² X 100,000) = 898,755,178,736.817,64 ESU of Capacitance
Henrys = 1/(299,792,458²/100,000) = 1.112,650,056,053,62E-12 ESU of Inductance
Coulombs = 1/(1/299,792,458)/10 = 2,997,924,580 ESU of Quantity

(Continued from page 173)

This is another one of the rather interesting results of this line of research and other recent quantum experiments.

The primordial singularity represents the "real" physical existence of this indivisible whole, and the real "physical substance" of which it consists is what is continuing to expand and spread out in the other "Dimensions" (existences) within the realm of "basic reality" (see page 136). According to this theory, no part of our universe has any existence outside of the sphere of influence of the primordial singularity ("physical substance") which also represents our total universe.

If and when the "physical substance" stops expanding and a wave function collapse of our universe does occur, one possible outcome would be that the "physical substance" might just possibly "explode" and start the process all over again. Another possible outcome may be that the "physical substance" which is expanding and spreading out (in other existences, or "Dimensions") and producing the universe that we see, may never stop expanding, and continue into infinity.

These are some of the possible outcomes, but it is probably unlikely that we will ever gain enough knowledge regarding the "physical substance" or the "void" which it is expanding into (realm of "basic reality"), to make anything better than an educated guess as to when and if a collapse may actually occur and then what will actually be the future of our universe.

There are some experiments which could be conducted to give us a better understanding of the physical characteristics (metrics) of the expansion of our universe, and which might possibly shed some light on what might happen in the future.

The first of these experiments would be to set up a very precise laser in a controlled environment (constant temperature, humidity, etc.), in which the beam would be reflected back and forth between a set of perfect mirrors in order to simulate a very large distance. Then we would need to determine, very precisely, the difference between the diameter of the laser beam as it is leaving the output lens of the laser, and the diameter of the of the laser beam as it would appear at a relatively large distance as simulated by the mirrors.

This size difference would probably give us a fairly accurate indication of the "rate" at which our universal space-time continuum is "stretching out" (changing Permittivity and Permeability values) as the "physical substance" continues to expand and thin out. What we are looking for, however, is the "rate of change" in this size difference.

Since we know that the speed of light is constant, and that the distance between the mirrors (as we see it) is constant, then the only thing that would be changing, observably, in this experiment is time. The difference in the size of the laser beam represents the passage of time (entropy), and any change in the rate at which time passes, however small, would most likely show up as a change in the difference in the size of the laser beam.

This theory predicts that the difference in size of the laser beam should change at the rate described by the curve represented by the "time line" (see page 71).

Another experiment we could perform would be to closely monitor the "gravitational constant", "Planck's constant", the "Permittivity constant", and the "Permeability constant" which are all closely related to the "pressure" of the physical medium.

If the rate at which the "pressure" or concentration of the physical medium decreases is slowing down gradually, then we might possibly see some slight changes in these values over a period of, say, 25 - 50 years.

If this data could be collected and plotted on a graph, any changes in these values might give us a better understanding of how the dissipation of the physical medium is related to the physical expansion rate of our universe. This information might shed some light on "where we are" on the graphs on page 15. If the "pressure" of the physical medium is still going down slightly, then our universe may have a considerable amount of time left (as we see it) before the physical medium reaches its maximum amount of dissipation.

If the "pressure" reduction rate of the physical medium is totally flat, then the physical medium could be nearing its maximum level of dissipation. Keep in mind, however, that due to the great size and age of our universe, any changes in the values mentioned above would be expected to be very, very small.

There is, however, another experiment which could help us to better-determine what our universe is going to do in the future. That would be to go ahead and build the "resonator" type of spacecraft described in chapter 6 and then use it to time-travel robots into the future and determine what our future holds.

The way to accomplish this would be to use this spacecraft to time-travel far back into the past during the time period when our early universe was just starting to form. Then, it could wait there for a pre-determined brief time period until our universe (where we are now) has continued to expand for many years into the future. The spacecraft could then time-travel back to the "where we are now" point on the "time chart" which, by that time, would have advanced many years into the future.

Once the observational, informational, and experimental results are collected and recorded, then the spacecraft could return back to our present time period and give us some idea what was going on during that particular future time period. Again, it is highly recommended that some type of robotic explorers would be used for such a dangerous mission which could have a very uncertain outcome.

As we have seen in Chapter 5, our universe is a closely integrated system which consists entirely of pressure-related distortions and velocity-related distortions in the primordial singularity as it exists in the realm of "basic reality". Some of the most convincing evidence we have to date which proves this statement, is the fact that electromagnetic waves can have both a transverse wave polarity as well as a longitudinal wave polarity.

Additionally, a closer examination and comparison of the graphs on page 15, page 71, and page 111, will demonstrate that the equations which these graphs represent will actually plot the same curve, and the shape of this curve is also quite similar to ¼ cycle of a sine wave which is a two-dimensional representation of an electromagnetic wave.

If we use this approach, we can use a process known as "numerical analysis" to gain a much better understanding of how pressure-related distortions in four-dimensional space-time, the amplitude of an electromagnetic wave which represents a quantum of energy and remains in a "fixed" state, for instance, are experiencing the effects of relativity (velocity-related space-time distortion) in order to produce what we perceive as: the shape of an electromagnetic wave (sine wave); objects with relative motion; gravity; mass; and energy.

By conducting these experiments, we can hope to gain a much better understanding of the relationship between the realm of "basic reality" and the four-dimensional space-time energy field (Permittivity and Permeability values) which produces the universe which we see.

With a better understanding of this underlying relationship, we will be in a much better position to fully interpret the rather unusual results of quantum experiments including: wave/particles which change physical states as a result of observations, and electromagnetic events which are instantaneous across large distances -- and how these experimental results can be interpreted to determine the future course which our universe will most likely take.

Concerning simplicity in regard to this theory, all you have is the interaction between the laws of physics in the "empty black void" and the characteristics of the "lumeniferous ether" and its effects. According to this theory, the interactions between the "empty black void" and the "lumeniferous ether" are able to produce our entire universe including the atoms, time, space, electric fields, magnetic fields, gravitational fields, and everything that we perceive.

The information in this book has been presented in such a manner as to challenge those who are attempting to understand the underlying physical processes which are producing the fascinating effects of quantum physics. Perhaps this information may be helpful in opening wider the door to our final frontier.

GLOSSARY

α ----- fine structure constant Introduced in 1916 by Arnold Sommerfeld (1868 – 1951) it represents the strength of the electromagnetic interaction between elementary charged particles.

abstract Considered apart from matter; imaginary. The essence of a larger object or whole.

astronomical Very large, as the numbers or quantities used in astronomy.

asymmetrical Not symmetrical, or, un-symmetrical. (see symmetrical)

axis A line around which a turning body rotates. A fixed line, as in a graph, along which distances are measured or to which positions are referred.

"basic reality" The "empty black void" in which our universe exists. There is no four-dimensional space-time energy field within this realm (except that which is contained within our universe), so the concepts of space and time, as we understand them, are meaningless beyond the outer reaches of our universe. For purposes of clarification, this realm was named "basic reality" because it represents what is actually real within our universe. Our perceptions of space and time are a grand-scale optical illusion which is being produced by our four-dimensional space-time energy field.

binary star system A pair of stars which orbit around a common center of gravity. Astronomers believe that most stars exist in this manner.

black hole A theoretical entity formed by the explosion (supernova) and resulting contraction of a large star, greater than 2 ½ times the mass of the sun. A region of space where the effects of the four-dimensional space-time energy field are almost totally distorted (gravitational field) by the tremendous amount of mass within the singularity which lies at the center of the black hole. According to this theory, <u>total</u> space-time distortion could be achieved by a black hole only if the total mass of the entire universe was once again contained within its singularity.

blackbody An ideal physical body which would absorb all incident radiation falling on it including all the radiation it emits.

bulk modulus The amount of force per unit area required to achieve a given amount of deformation.

chronometer Precision clock for accurate measurements of time over long durations.

cosmology (as it applies to astronomy and astrophysics) is the study of the structure, dynamics, and the development of our universe. A cosmology theory attempts to explain how our universe was formed, what has happened to it in the past, and what might happen to it in the future.

Doppler effect The change, or shift, in the frequency of sound or light (electromagnetic) waves caused by the relative motion of the source of the waves either toward or away from the observer.

dynamic   Active; Having to do with physical energy; Constantly changing;

effects   (of four-dimensional space-time) Impressions that we have about our universe and what we see around us. Relativity is an accurate description of the effects of four-dimensional space-time and quantum theory is an accurate description of the interaction between the effects of four-dimensional space-time and the realm of "basic reality".

elasticity   Spontaneously returning to former shape after compression, extension, or other distortion. The property or quality of being elastic.

electromagnetic wave polarization   The orderly alignment of the electric and magnetic fields of which an electromagnetic wave/particle consists. Once the electric and magnetic fields are aligned in a particular orientation, they will be held in that state by the nature of the void "basic reality" in which our universe exists.

entropy   The principle , most commonly applied in Thermodynamics, which states that the complexity or disorder in a system increases as time proceeds.

EPR paradox   Written in 1935 by Albert Einstein, Boris Podolsky, and Nathan Rosen, it represented a hypothetical challenge to Quantum Mechanics that electromagnetic wave/particles could communicate with each other instantly across arbitrarily large distances of space – faster than the speed of light.

essence   That which makes something what it appears to be.

format   The general form or arrangement of anything.

four-dimensional space-time   In this theory, it is referred to as the abstract energy field which produces our perceptions (what we see) regarding the universe in which we live. It is represented by two electrical terms known as the Permittivity of free space and the Permeability of free space (see page 30).

galaxy   A large cluster of stars, gas, dust, and, in many cases, a black hole.

homogeneous   Of the same kind, order, or nature. All the same.

inertial reference frame   A reference frame in which an object experiences no net force due to other bodies. A spacecraft travelling in a straight line at a constant velocity, not rotating, and far from any gravitational fields, would be an example.

inverse   A reversed or opposite orientation. As one value becomes larger, the other becomes proportionately smaller and vice versa.

isotropic   Evenly spread. Evenly distributed throughout.

kinetic   Of, or pertaining to motion. Kinetic energy; energy of motion; energy in motion.

lumeniferous ether   Early physicists believed that this was the physical substance through which electromagnetic waves propagate (travel). In this theory, there actually is a physical medium which produces our perceptions of electromagnetic waves, but it has a "real" existence only within the primordial singularity within our "Dimension" (universe) (see page 136). The actual physical characteristics of this substance are represented by the Permittivity and Permeability of free space.

molecule   Two or more atoms joined together.

negative space-time distortion   The space and time relationship which would be experienced by two objects which have a relative motion away from each other, and an object in a gravitational field. The Permittivity and Permeability values would be reduced.

paradox   A statement which seems to contradict itself, but which may be true.

Permeability of free space   A value of proportionality that exists between magnetic flux density and magnetic field strength. (μ_0) the Permeability of free space: 1.256,637,061,44E-7 Henry's per meter^{-1} (The $^{-1}$ exponent means that there is an inverse relationship between ε_0 and μ_0)

Permittivity of free space   A value of proportionality that exists between electric flux density and electric field intensity. (ε_0) the Permittivity of free space: (8.854,187E-12 Farads per meter)

perpendicular   At a right angle to.

point of "basic reality"   The point at which our concepts of space and time become meaningless. A black hole singularity, a neutron, or something travelling at the speed of light all represent a reference frame in which the effects of the four-dimensional space-time energy field are distorted down to very nearly zero, or 1.65×10^{-13} cm.

positive space-time distortion   The space and time relationship which would be experienced by two objects which have a relative motion <u>toward</u> each other. The Permittivity and Permeability values would be increased.

primordial singularity   The small (1.65×10^{-13} cm), highly compressed singularity which is held in a "fixed state" within the realm of "basic reality". The singularity has actually expanded in other dimensions, or existences, to create the four-dimensional space-time energy field which produces our universe that we see today.

propagate   Travel through, as in wave motion.

proportionately   Symmetrical in relative magnitude or arrangement. Equality of ratios.

quanta   Discrete amounts of electromagnetic wave energy or energy states associated with atomic interactions. Electromagnetic wavelets carrying energy equal to their amplitude multiplied by their frequency.

quantum entanglement   A phenomenon in which two or more separate electromagnetic wave/particles are interactive and instantaneous across arbitrarily large distances of space. If a measurement is made on one of the wave/particles, the other(s) will "know" instantly what measurement has been made and also the outcome of that measurement.

quantum event   Any electromagnetic disturbance which involves waves in the four-dimensional space-time energy field (represented by the Permittivity and Permeability of free space).

solar system   The Sun, including the planets, with orbiting moons,

spectrum   An orderly display of electromagnetic radiation which has been diffracted and arranged in the order of the respective wavelengths, or frequencies, of which the radiation consists.

sphere   A geometrical form in which all points on the surface are equidistant from the center point.   In this theory, the center point of a sphere can also represent all points on the surface, and all points within the sphere simultaneously (quantum mechanics).

"sphere of influence"   The space component, or wave function (as we see it) of an electromagnetic wave as it expands outward in all directions at the speed of light.

spring constant   The measure of "stiffness" or "elasticity" within an object, a particular substance, or medium which is capable of producing simple harmonic motion.

stellar   of the stars, or of a star

supernova   A star which contains at least 2½ times the mass of our Sun, and which burns its hydrogen and helium much more rapidly than a less-massive star due to its increased gravitational pressure.   Its burning process continues to produce heavier and heavier elements until it starts producing significant quantities of uranium. It first collapses, and then it undergoes a huge explosion in which the core can be ripped apart to form planets and moons, or it can condense even further to form a black hole.

symmetrical   A condition in which one part is proportionate to another. Composed of two parts which correspond harmoniously to each other.

viscosity   A measure of a fluid's resistance to flow or movement which may also be referred to as "thickness".

wave/particle   An electromagnetic wave in four-dimensional space-time which propagates through space at the speed of light. It can also appear as a particle due to its existence within the primordial singularity at the point of "basic reality".

BIBLIOGRAPHY

These are some of the books you may consider reading which provide informational support and to help you envision the bigger picture so you can see how all this information fits together. By using this theory as an outline, you can have many wonderful hours of interesting reading while piecing together the observations and experimental results which are presented clearly and concisely in each of these great works.

Alexandrov, Yu. A.. **Fundamental Properties of the Neutron.** Oxford: Oxford University Press, 1992.

Boslough, John. **Masters of Time.** New York: Addison-Wesley, 1992.

Boslough, John, "Searching for the Secrets of Gravity". National Geographic, 175:5, May 1989.

Calder, Nigel. **Einstein's Universe.** New York: Greenwich House, 1979.

Cohen-Tannoudji, Gilles. **Universal Constants in Physics.** New York: McGraw-Hill, Inc., 1991.

Coveney, Peter, and Roger Highfield. **The Arrow of Time.** New York: Fawcett Columbine, 1990.

Davies, Paul, and Gribbin, John. **The Matter Myth.** New York: Simon & Shuster, 1992.

Dirac, P. A. M. **The Principles of Quantum Mechanics.** New York: Oxford University Press, 1958.

Einstein, Albert. **RELATIVITY The Special and the General Theory.** New York: Crown Publishers, Inc., 1961.

Einstein, Albert; Lorentz, H. A.; Minkowski, H.; and Weyl, H. **The Principle of Relativity.** New York: Dover Publications, Inc., 1952.

Ellis, George F. R., and Williams, Ruth M. **Flat and Curved Space-Times.** New York: Oxford University Press, 1988.

Gazdag, László. **Beyond the Theory of the Relativity**. Budapest: Ernő Werőczei, 1998.

Giancoli, Douglas C. **General Physics.** Englewood Cliffs, New Jersey: Prentice-Hall, 1984.

Gribbin, John. **In Search of Schrödinger's Cat.** New York: Bantam, 1984.

Hawking, Stephen W. **A Brief History of Time.** New York: Bantam, 1988.

Herbert, Nick. **Quantum Reality.** New York: Anchor Books, 1985.

Horgan, John. "**Quantum Philosophy**". Scientific American, July, 1990.

Kaufman, Nicholas J. **Black Holes and Warped Space-time.** W. H. Freeman and Company, 1979.

Macrae, Norman. **John von Neumann.** New York: Pantheon Books, 1992.

Marion, Jerry B. **Our Physical Universe.** New York: John Wiley & Sons, Inc., 1971.

Martin, Thomas C.. **The Inventions, Researches, and Writings of Nikola Tesla.** New York: Barnes & Noble Books, 1992.

Matlin, Marlee; Hendrix, Elaine; Bailey, Robert, Jr.. and numerous other authors. **The Little Book of BLEEPS.** USA: Captured Light Distribution, 2004.

Maxwell, James Clerk. **A Treatise on Electricity and Magnetism.** Oxford: Clarendon Press, 1891.

Moore, William L. in consultation with Berlitz, Charles. **The Philadelphia Experiment:.** New York: Grosset & Dunlap Publishers, 1979.

Murdin, Paul. **End in Fire.** New York: Cambridge University Press, 1990.

Pais, Abraham. **Niels Bohr's Times, in Physics, Philosophy, and Polity.** Oxford: Oxford University Press, 1991.

Peat, F. David. **Einstein's Moon.** Chicago: Contemporary Books, 1990.

Pierls, Rudolph. **More Surprises in Theoretical Physics.** Princeton, New Jersey: Princeton University Press, 1991.

Schrödinger, Erwin. **Space-Time Structure.** New York: Cambridge University Press, 1950.

APPENDIX

Twelve Star Publishing : Sourceworks

Ether and the Theory of Relativity

(An address delivered by Dr. Albert Einstein on May 5th, 1920, in the University of Leyden, Netherlands)

How does it come about that alongside of the idea of ponderable matter, which is derived by abstraction from everyday life, the physicists set the idea of the existence of another kind of matter, the ether? The explanation is probably to be sought in those phenomena which have given rise to the theory of action-at-a-distance, and in the properties of light which have led to the undulatory theory. Let us devote a little while to the consideration of these two subjects.

Outside of physics we know nothing of action-at-a-distance. When we try to connect cause and effect in the experiences which natural objects afford us, it seems at first as if there were no other mutual actions than those of immediate contact, e.g. the communication of motion by impact, push and pull, heating or inducing combustion by means of a flame, etc. It is true that even in everyday experience weight, which is in a sense action-at-a-distance, plays a very important part. But since in daily experience the weight of bodies meets us as something constant, something not linked to any cause which is variable in time or place, we do not in everyday life speculate as to the cause of gravity, and therefore do not become conscious of its character as action-at-a-distance. It was Newton's theory of gravitation that first assigned a cause for gravity by interpreting it as action-at-a-distance, proceeding from masses. Newton's theory is probably the greatest stride ever made in the effort towards the casual nexus of natural phenomena. And yet this theory evoked a lively sense of discomfort among Newton's contemporaries, because it seemed to be in conflict with the principle springing from the rest of experience, that there can be reciprocal action only through contact, and not through immediate action-at-a-distance.

It is only with reluctance that man's desire for knowledge endures a dualism of this kind. How was unity to be presented in his comprehension of the forces of nature? Either by trying to look upon contact forces as being themselves distant forces which admittedly are observable only at a very small distance—and this was the road which Newton's followers, who were entirely under the spell of his doctrine, mostly preferred to take; or by assuming that the Newtonian action-at-a-distance is only *apparently* immediate action-at-a-distance, but in truth is conveyed by a medium permeating space, whether by movements or by elastic deformation of this medium. Thus the endeavor toward a unified view of the nature of forces leads us to the hypothesis of an ether. This hypothesis, to be sure, did not at first bring with it any advance in the theory of gravitation or physics generally, so that it became customary to treat Newton's law of force as an axiom not further reducible.

But the ether hypothesis was bound always to play some part in physical science, even if at first only a latent part.

When in the first half of the nineteenth century the far-reaching similarity was revealed which subsists between the properties of light and those of elastic waves in ponderable bodies, the ether hypothesis found fresh support. It appeared beyond question that light must be interpreted as a vibratory process in an elastic, inert medium filling up universal space. It also seemed to be a necessary consequence of the fact that light is capable of polarization, that this medium, the ether, must be of the nature of a solid body, because transverse waves are not possible in a fluid, but only in a solid. Thus the physicists were bound to arrive at the theory of the "quasi-rigid" lumeniferous ether, the parts of which can carry out no movements relatively to one another except the small movements of deformation which correspond to light-waves.

This theory – also called the theory of the stationary lumeniferous ether – moreover found a strong support in an experiment which is also of fundamental importance in the special theory of relativity, the experiment of Fizeau, from which one was obliged to infer that the lumeniferous ether does not take part in the movements of bodies. The phenomenon of aberration also favoured the theory of the quasi-rigid ether.

The development of the theory of electricity along the path opened up by Maxwell and Lorentz gave the development of our ideas concerning the ether quite a peculiar and unexpected turn. For Maxwell himself, the ether indeed still had properties which were purely mechanical although of a much more complicated kind than the mechanical properties of tangible solid bodies. But neither Maxwell nor his followers succeeded in elaborating a mechanical model for the ether which might furnish a satisfactory mechanical interpretation of Maxwell's laws of the electromagnetic field. The laws were clear and simple, the mechanical interpretations clumsy and contradictory. Almost imperceptibly the theoretical physicists adapted themselves to a situation which, from the standpoint of their mechanical programme, was very depressing. They were particularly influenced by the electro-dynamical investigations of Heinrich Hertz. For whereas they previously had required of a conclusive theory that it should content itself with the fundamental concepts which belong exclusively to mechanics (e.g. densities, velocities, deformations, stresses) they gradually accustomed themselves to admitting electric and magnetic force as fundamental concepts side by side with those of mechanics, without requiring a mechanical interpretation for them. Thus the purely mechanical view of nature was gradually abandoned.

But this change led to a fundamental dualism which in the long-run was unsupportable. A way of escape was now sought in the reverse direction, by reducing the principles of mechanics to those of electricity, and this is especially as confidence in the strict validity of the equation of Newton's mechanics was shaken by the experiments with *beta*-rays and rapid cathode rays.

This dualism still confronts us in unextenuated form in the theory of Hertz, where matter appears not only as the bearer of velocities, kinetic energy, and mechanical pressures, but also as the bearer of electromagnetic fields. Since such fields also occur *in vacuo* – i.e. in free ether – the ether also appears as bearer of electromagnetic fields.

The ether appears indistinguishable in its functions from ordinary matter. Within matter, it takes part in the motion of matter, and in empty space it has everywhere a velocity; so that the ether has a definitely assigned velocity throughout the whole of space. There is no fundamental difference between Hertz's ether and ponderable matter (which in part subsists in the ether).

The Hertz theory suffered not only from the defect of ascribing to matter and ether, on one hand mechanical states, and on the other hand electrical states, which do not stand in any conceivable relation to each other, it was also at variance with the result of Fizeau's important experiment on the velocity of the propagation of light in moving fluids, and with other established experimental results.

Such was the state of things when H. A. Lorentz entered upon the scene. He brought theory into harmony with experience by means of a wonderful simplification of theoretical principles. He achieved this, the most important advance in the theory of electricity since Maxwell, by taking from the ether its mechanical, and from matter its electromagnetic qualities. As in empty space, so too in the interior of material bodies, the ether, and not matter viewed atomistically, was exclusively the seat of electromagnetic fields. According to Lorentz the elementary particles of matter alone are capable of carrying out movements; their electromagnetic activity is entirely confined to the carrying of electrical charges. Thus Lorentz succeeded in reducing all electromagnetic happenings to Maxwell's equations for free space.

As to the mechanical nature of the Lorentzian ether, it may be said of it, in a somewhat playful spirit, that immobility is the only mechanical property of which it has not been deprived by H. A. Lorentz. It may be added that the whole change in the conception of the ether which the special theory of relativity brought about, consisted in taking away from the ether its last mechanical quality, namely, its immobility. How this is to be understood will forthwith be expounded.

The space-time theory and the kinematics of the special theory of relativity were modelled on the Maxwell-Lorentz theory of the electromagnetic field. This theory therefore satisfies the conditions of the special theory of relativity, but when viewed from the latter it acquires a novel aspect. For if *K* be a system of co-ordinates relatively to which the Lorentzian ether is at rest, the Maxwell-Lorentz equations are valid permanently with reference to K. But by the special theory of relativity, the same equations, without any change of meaning, also hold in relation to any new system of co-ordinates K' which is moving in uniform translation relatively to K. Now comes the anxious question: -- Why must I in the theory distinguish the K system above all K' systems, which are physically equivalent to it in all respects, by assuming that the ether is at rest relatively to the K system? For the theoretician such an asymmetry in the theoretical structure, with no corresponding asymmetry in the system of experience, is intolerable. If we assume the ether to be at rest relatively to K, but in motion relatively to K', the physical equivalence of *K* and *K'* seems to me from the logical standpoint, not indeed downright incorrect, but nevertheless inacceptable.

The next position which it was possible to take up in the face of this state of things appeared to be the following. The ether does not exist at all. The electromagnetic fields are not states of a medium, and are not bound down to any bearer, but they are independent realities which are not reducible to anything else, exactly like the atoms of ponderable matter. This conception suggests itself the more readily as, according to Lorentz's theory, electromagnetic radiation, like ponderable matter, brings impulse and energy with it, and as, according to the special theory of relativity, both matter and radiation are but special forms of distributed energy, ponderable mass losing its isolation and appearing as a special form of energy.

More careful reflection teaches us, however, that the special theory of relativity does not compel us to deny the ether. We may assume the existence of an ether, only we must give up ascribing a definite state of motion to it, i.e. we must by abstraction take from it the last mechanical characteristic which Lorentz had still left it. We shall see later that this point of view, the conceivability of which I shall at once endeavor to make more intelligible by a somewhat halting comparison, is justified by the results of the general theory of relativity.

Think of waves on the surface of water. Here we can describe two entirely different things. Either we may observe how the undulatory surface forming the boundary between water and air alters in the course of time; or else – with the help of small floats, for instance – we can observe how the position of the separate particles of water alters during the course of time. If the existence of such floats for tracking the motion of the particles of a fluid were a fundamental impossibility in physics – if, in fact, nothing else whatever were observable than the shape of the space occupied by the water as it varies in time, we should have no ground for the assumption that water consists of moveable particles. But all the same we could characterize it as a medium.

We have something like this in the electromagnetic field. For we may picture the field to ourselves as consisting of lines of force. If we wish to interpret these lines of force to ourselves as something material in the ordinary sense, we are tempted to interpret the dynamic processes as motions of these lines of force, such that each separate line of force is tracked through the course of time. It is well known, however, that this way of regarding the electromagnetic field leads to contradictions.

Generalizing we must say this: -- There may be supposed to be extended physical objects to which the idea of motion cannot be applied. They may not be thought of as consisting of particles which allow themselves to be separately tracked through time. In Minkowski's idiom this is expressed as follows: -- Not every extended conformation in the four dimensional world can be regarded as composed of world threads. The special theory of relativity forbids us to assume the ether to consist of particles observable through time, but the hypothesis of ether in itself is "not" in conflict with the special theory of relativity. Only we must be on our guard against ascribing a state of motion to the ether.

Certainly, from the standpoint of the special theory of relativity, the ether hypothesis appears at first to be an empty hypothesis. In the equations of the electromagnetic field there occur, in addition to the densities of the electric charge, *only* the intensities of the field. The career of electromagnetic processes *in vacuo* appears to be completely determined by (these) equations, uninfluenced by other physical quantities. The electromagnetic fields appear as ultimate, irreducible realities, and at first it seems superfluous to postulate a homogeneous, isotropic ether-medium, and to envisage electromagnetic fields as states of this medium.

But on the other hand there is a weighty argument to be adduced in favor of the ether hypothesis. To deny the ether is ultimately to assume that empty space has no physical qualities whatever. The fundamental facts of mechanics do not harmonize with this view. For the mechanical behavior of a corporeal system hovering freely in empty space depends not only on relative positions (distances) and relative velocities, but also on its state of rotation, which physically may be taken as a characteristic not appertaining to the system itself. In order to be able to look upon the rotation of the system, at least formally, as something real, Newton objectivises space. Since he classes his absolute space together with real things, for him rotation relative to an absolute space is also something real. Newton might no less well have called his absolute space "Ether"; what is essential is merely that besides observable objects, another thing, which is not perceptible, (must) be looked upon as real, to enable acceleration or rotation to be looked upon as something real.

It is true that Mach tried to avoid having to accept as real something which is not observable by endeavoring to substitute in mechanics a mean acceleration with reference to the totality of the masses in the universe in place of an acceleration with reference to absolute space. But inertial resistance opposed to relative acceleration of distant masses presupposes action-at-a-distance; and as the modern physicist does not believe that he may accept this action-at-a-distance, he comes back once more, if he follows Mach, to the ether, which has to serve as (the) medium for the effects of inertia. But this conception of the ether to which we are led by Mach's way of thinking differs essentially from the ether as conceived by Newton, by Fresnel, and by Lorentz. Mach's ether not only *conditions* the behavior of inert masses, but is *also conditioned* in its state by them.

Mach's idea finds its full development in the ether of the general theory of relativity. According to this theory the metrical qualities of the continuum of space-time differ in the environment of different points of space-time, and are partly conditioned by the matter existing outside of the territory under consideration. This space-time variability of the reciprocal relations of the standards of space and time, or, perhaps, the recognition of the fact that empty space in its physical relation is neither homogenous nor isotropic compelling us to describe its state by ten functions (the gravitational potentials g, *mu, nu*), has, I think, finally disposed of the view that space is physically empty. But therewith the conception of the ether has again acquired an intelligible content, although this content differs widely from that of the ether of the mechanical undulatory theory of light. The ether of the general theory of relativity is a medium which is itself devoid of all mechanical and kinematical qualities, but helps to determine mechanical (and electromagnetic) events.

What is fundamentally new in the ether of the general theory of relativity as opposed to the ether of Lorentz consists in this, that the state of the former is at every place determined by connections with matter and the state of the ether in neighboring places, which are amenable to law in the form of differential equations; whereas the state of the Lorentzian ether in the absence of electromagnetic fields is conditioned by nothing outside itself, and is everywhere the same. The ether of the general theory of relativity is transmuted conceptually into the ether of Lorentz if we substitute constants for the functions of space which describe the former, disregarding the causes which condition its state. Thus we may also say, I think, that the ether of the general theory of relativity is the outcome of the Lorentzian ether, through relativation.

As to the part which the new (conception of the ether) is to play in the physics of the future we are not yet clear. We know that it determines the metrical relations in the space-time continuum, e.g. the configurative possibilities of solid bodies as well as the gravitational fields, but we do not know whether it has an essential share in the structure of the electrical elementary particles constituting matter. Nor do we know whether it is only in the proximity of ponderable masses that its structure differs essentially from that of the Lorentzian ether; whether the geometry of spaces of cosmic extent is approximately Euclidean. But we can assert by reason of the gravitation that there must be a departure from Euclidean relations, with spaces of cosmic order of magnitude if there exists a positive mean density, no matter how small, of the matter in the universe. In this case the universe must of necessity be spatially unbounded and of finite magnitude, its (magnitude) being determined by the value of that mean density.

If we consider the gravitational field and the electromagnetic field from the standpoint of the ether hypothesis, we find a remarkable difference between the two. There can be no space nor any part of space without gravitational potentials; for these confer upon space its metrical qualities, without which it cannot be imagined at all. The existence of the gravitational field is inseparably bound up with the existence of space. On the other hand a part of space may be very well imagined without an electromagnetic field; thus in contrast with the gravitational field, the electromagnetic field seems to be only secondarily linked to the ether, the formal nature of the electromagnetic field being as yet in no way determined by that of gravitational ether. From the present state of theory it looks as if the electromagnetic field, as opposed to the gravitational field, rests upon an entirely new formal *motif*, as though nature might just as well have endowed the gravitational ether with fields of quite another type, for example, with fields of a scalar potential, instead of fields of the electromagnetic type.

Since according to our present conceptions the elementary particles of matter are also, in their essence, (are) nothing else than condensations of the electromagnetic field. Our present view of the universe presents two realities which are completely separated from each other conceptually, although connected causally, namely, gravitational ether and electromagnetic field, or, as they might also be called(,) space and matter.

Of course it would be a great advance if we could succeed in comprehending the gravitational field and the electromagnetic field together as one unified conformation. Then for the first time the epoch of theoretical physics founded by Faraday and Maxwell would reach a satisfactory conclusion. The contrast between the ether and matter would fade away, and, through the general theory of relativity, the whole of physics would become a complete system of thought, like geometry, kinematics, and the theory of gravitation. An exceedingly ingenious attempt in this direction has been made by the mathematician H. Weyl, but I do not believe that his theory will hold its ground in relation to reality. Further, in contemplating the immediate future of theoretical physics, we ought not unconditionally to reject the possibility that the facts comprised in the quantum theory may set bounds to the field theory beyond which it cannot pass.

Recapitulating, we may say that according to the general theory of relativity, space is endowed with physical qualities, in this sense, therefore, there exists an ether.

According to the general theory of relativity space without ether is unthinkable. For in such space there would be no propagation of light, but also no possibility of existence for standards of space and time (measuring-rods and clocks), not therefore, any space-time intervals in the physical sense. But this ether may not be thought of as endowed with the quality characteristic of ponderable media, as consisting of parts which may be tracked through time. The idea of motion may not be applied to it.

HIGH-PRECISION TABLE OF CONSTANTS

Copyright © 2015 ALL RIGHTS RESERVED

The physical substance of which the singularity consists appears to have characteristics normally associated with a solid material since it is most likely being held in a fixed state due to the absence of scalar properties (space-time) in the void in which it exists. These conditions provide the basic structure for our physical and numerical constants. The six basic constants: c; e^2; α; $\hbar$; ε_0; and μ_0 appear to represent the basic internal structure and state of the physical substance within the primordial singularity immediately prior to its expansion, and therefore, may be considered as dimensionless during this stage of the process.

At the instant in which the singularity began to expand, the scalar properties of space, time, acceleration, and relative motion/heat came into existence which added increasing levels of complexity (thermal entropy) to the values which the higher-level constants represent -- with the six basic constants still representing the underlying structure from which the higher-level constants are patterned. The Permittivity (ε_0) and Permeability (μ_0) of free space then represent these quantities into our reality (our perception) in their respective proportions and forms which include: meters per second; meters per second^2; Coulombs of electric charge; Farads of capacitance; Henry's of inductance; etc. In some cases, the exponents and units have been intentionally omitted since these values may be numerically representing several different quantities.

In many of the following examples, the calculated values of the physical constants may vary from the experimentally measured values by a few decimal places. Consequently, the primary objective here is to suggest the existence of an underlying physical framework which will ultimately provide us with the means of accurately calculating the precise values of all physical constants. The values of the physical constants which are represented here are to be used as examples only. **They are not intended to replace the currently established values which have been determined by experiment unless reviewed and accepted by the scientific community as a whole. Prior to utilizing any of the following terms and formulas, a math re-check should be conducted to check for errors.**

By setting the speed of light to exactly c = 299,792,458 meters per second, the precise values of ε_0 and μ_0 are now established as a metric standard by which all of the other physical constants may be measured. The precise values of each of the six basic constants may now be determined to a high degree of accuracy by definition which, as a result, enables us to calculate all of the other physical constants to a high level of accuracy.

Physical orientations of the six basic constants:

$$c = \frac{\alpha \hbar}{e^2} \,;\quad e^2 = \frac{\alpha \hbar}{c} \,;\quad \alpha = \frac{c e^2}{\hbar} \,;\quad \hbar = \frac{c e^2}{\alpha} \,;\quad \frac{e^2}{\alpha \hbar \mu_0} = \frac{\alpha \hbar \varepsilon_0}{e^2} \,;$$

Permittivity of free space:

ε_0 = **8.854,187,817,620,389,850,536,563,031,710,9e-12 Farads/meter**

$$\varepsilon_0 = \frac{e^4}{\alpha^2 \hbar^2 \mu_0} = \frac{e^2}{\alpha \hbar c \mu_0} = \frac{1}{c^2 \mu_0} = \frac{2}{8\pi c^2}$$

$$\varepsilon_0^2 = \frac{e^8}{\alpha^4 \hbar^4 \mu_0^2} = \frac{1}{c^4 \mu_0^2} = 7.839{,}664{,}190{,}969{,}732{,}200{,}220{,}745{,}818{,}350{,}1\text{e-}23 \text{ Farads/meter}$$

$$2\varepsilon_0 = \frac{1}{2\pi c^2} = \frac{e^4}{2\alpha^2 \hbar^2 \pi} = 1.770{,}837{,}563{,}524{,}077{,}970{,}107{,}312{,}606{,}342{,}2$$

$$4\varepsilon_0 = \frac{1}{\pi c^2} = \frac{e^4}{\alpha^2 \hbar^2 \pi} = 3.541{,}675{,}127{,}048{,}155{,}940{,}214{,}625{,}212{,}684{,}4$$

$$c^2 \varepsilon_0 = \frac{\alpha^2 \hbar^2 \varepsilon_0}{e^4} = \frac{1}{\mu_0} = 7.957{,}747{,}154{,}594{,}766{,}788{,}444{,}188{,}168{,}625{,}8$$

$$\frac{1}{\varepsilon_0 c} = \frac{e^2}{\alpha \hbar \varepsilon_0} = \mu_0 c = \; = 3.767{,}303{,}134{,}617{,}706{,}554{,}681{,}984{,}004{,}203{,}2$$

$$\sqrt{\varepsilon_0} = \frac{e^2}{\alpha \hbar \sqrt{\mu_0}} = 2.975{,}598{,}732{,}628{,}509{,}144{,}381{,}087{,}753{,}323{,}5\text{e-}6 \text{ Farads/meter}$$

Permeability of free space:

μ_0 = **12.566,370,614,359,172,953,850,573,533,118e-7 Henry's/meter^{-1}**

$$\mu_0 = 4\pi = \frac{e^4}{\alpha^2 \hbar^2 \varepsilon_0} = \frac{e^2}{\alpha \hbar c \varepsilon_0} = \frac{1}{\varepsilon_0 c^2} = \frac{2h}{\hbar}$$

$2\mu_0$ = 2.513,274,122,871,834,590,770,114,706,623,6e-6 Henry's/meter^{-1}

$$2\mu_0 = 8\pi = \frac{2}{c^2 \varepsilon_0} = \frac{1}{\sqrt{\pi \varepsilon_0} c^3 \varepsilon_0} = \frac{\sqrt{\pi \varepsilon_0}}{\pi c^3 \varepsilon_0^2}$$

$$c^2 \mu_0 = \frac{\alpha^2 \hbar^2 \mu_0}{e^4} = \frac{1}{\varepsilon_0} = 11{,}294{,}090{,}667.581{,}471{,}383{,}508{,}233{,}929{,}368$$

Orientation of the Permittivity and Permeability constants:

$$c\varepsilon_0 = \frac{1}{c\mu_0} = \frac{e^2}{\alpha\hbar\mu_0} = \frac{\alpha\hbar\varepsilon_0}{e^2} = 2.654,418,729,438,072,384,210,608,850,148,5$$

$$c\mu_0 = \frac{1}{c\varepsilon_0} = \frac{e^2}{\alpha\hbar\varepsilon_0} = \frac{\alpha\hbar\mu_0}{e^2} = 37.673,031,346,177,065,546,819,840,042,032$$

$$\frac{\varepsilon_0}{\sqrt{\pi\varepsilon_0}} = \frac{\sqrt{\pi\varepsilon_0}}{\pi} = = 5.308,837,458,876,144,768,421,217,700,297,1$$

Conductance quantum: G_0 = 7.748,091,734,636,263,793,631,291,999,104,2

$$G_0 = \frac{e^2}{\pi\hbar} = \frac{\alpha}{\pi c} = \frac{2e^2}{h} = \frac{e^2}{\sqrt{\pi\varepsilon_0}hc} = \frac{\sqrt{\pi\varepsilon_0}e^2}{h\pi c\varepsilon_0}$$

Speed of light:

$$c = \frac{\alpha\hbar}{e^2} = \frac{2h\alpha}{e^2\mu_0} = \frac{1}{\sqrt{\varepsilon_0\mu_0}} = \frac{1}{c\varepsilon_0\mu_0} = \textbf{299,792,458 meters per second}$$

$$\frac{1}{c} = \textbf{3.335,640,951,981,520,495,755,767,144,749,2e-9 meters per second}$$

$$\frac{1}{c} = \frac{e^2}{\alpha\hbar} = \frac{e^2\mu_0}{2h\alpha} = \sqrt{\varepsilon_0\mu_0} = c\varepsilon_0\mu_0 = 2\sqrt{\pi\varepsilon_0} = \frac{c}{c^2}$$

$$c^2 = \frac{\alpha^2\hbar^2}{e^4} = \frac{1}{\varepsilon_0\mu_0} = \textbf{89,875,517,873,681,764 meters per second}$$

$$\frac{1}{c^2} = \frac{e^4}{\alpha^2\hbar^2} = c^2(\varepsilon_0\mu_0)^2 = \varepsilon_0\mu_0 = \textbf{1.112,650,056,053,618,432,174,089,964,848e-17 m/s}$$

$$c^3 = \frac{\alpha^3\hbar^3}{e^6} = \frac{c}{\varepsilon_0\mu_0} = \textbf{26,944,002,417,373,989,539,335,912 m/s}$$

$$\frac{1}{c^3} = \frac{e^6}{\alpha^3\hbar^3} = \frac{\varepsilon_0\mu_0}{c} = \textbf{3.711,401,092,196,983,928,708,740,113,962,5e-26 m/s}$$

$$c^4 = \frac{\alpha^4 \hbar^4}{e^8} = \frac{1}{(\varepsilon_0 \mu_0)^2} = 80,776,087,130,624,902,292,638,00,461,517\text{e-}33 \text{ m/s}$$

$$\frac{1}{c^4} = \frac{e^8}{\alpha^4 \hbar^4} = (\varepsilon_0 \mu_0)^2 = 1.237,990,147,236,120,239,125,141,738,543,2\text{e-}34 \text{ m/s}$$

$$c^5 = \frac{\alpha^5 \hbar^5}{e^{10}} = \frac{c}{(\varepsilon_0 \mu_0)^2} = 24,216,061,708,512,206,534,319,783,561,111\text{e}42 \text{ m/s}$$

$$\frac{1}{c^5} = \frac{e^{10}}{\alpha^5 \hbar^5} = \frac{(\varepsilon_0 \mu_0)^2}{c} = 4.129,490,633,270,434,839,041,687,094,553,8\text{e-}43 \text{ m/s}$$

$$c^6 = \frac{\alpha^6 \hbar^6}{e^{12}} = \frac{1}{(\varepsilon_0 \mu_0)^3} = \frac{c^2}{(\varepsilon_0 \mu_0)^2} = 72,597,926,626,745,539,199,273,892,718,133\text{e}50 \text{ m/s}$$

$$\frac{1}{c^6} = \frac{e^{12}}{\alpha^6 \hbar^6} = (\varepsilon_0 \mu_0)^3 = \frac{(\varepsilon_0 \mu_0)^2}{c^2} = 1.377,449,806,716,096,520,027,093,908,597\text{e-}51 \text{ m/s}$$

$$c^7 = \frac{\alpha^7 \hbar^7}{e^{14}} = \frac{c}{(\varepsilon_0 \mu_0)^3} = 21,764,310,869,135,693,737,085,672,113,197\text{e}59 \text{ m/s}$$

$$\frac{1}{c^7} = \frac{e^{14}}{\alpha^7 \hbar^7} = \frac{(\varepsilon_0 \mu_0)^3}{c} = 4.594,677,984,581,241,600,237,634,759,303,6\text{e-}60 \text{ m/s}$$

$$c^8 = \frac{\alpha^8 \hbar^8}{e^{16}} = \frac{1}{(\varepsilon_0 \mu_0)^4} = \frac{c^2}{(\varepsilon_0 \mu_0)^3} = 65,247,762,521,343,059,609,761,193,993,975\text{e}36 \text{ m/s}$$

$$\frac{1}{c^8} = \frac{e^{16}}{\alpha^8 \hbar^8} = (\varepsilon_0 \mu_0)^4 = \frac{(\varepsilon_0 \mu_0)^3}{c^2} = 1.532,619,604,653,710,668,144,171,511,914\text{e-}36 \text{ m/s}$$

$$c^9 = \frac{\alpha^9 \hbar^9}{e^{18}} = \frac{c}{(\varepsilon_0 \mu_0)^4} = 19,560,787,105,273,713,301,650,829,140,469\text{e}45 \text{ m/s}$$

$$\frac{1}{c^9} = \frac{e^{18}}{\alpha^9 \hbar^9} = \frac{(\varepsilon_0 \mu_0)^4}{c} = 5.112,268,717,092,645,032,931,987,608,288,3\text{e-}77 \text{ m/s}$$

$$c^{10} = \frac{\alpha^{10} \hbar^{10}}{e^{20}} = \frac{1}{(\varepsilon_0 \mu_0)^5} = \frac{c^2}{(\varepsilon_0 \mu_0)^4} = 5.864,176,446,704,711,273,489,197,525,75\text{e}84 \text{ m/s}$$

$$\frac{1}{c^{10}} = \frac{e^{20}}{\alpha^{10} \hbar^{10}} = (\varepsilon_0 \mu_0)^5 = \frac{(\varepsilon_0 \mu_0)^4}{c^2} = 1.705,269,289,026,825,695,839,215,410,9\text{e-}85 \text{ m/s}$$

Elementary charge: $e = \sqrt{\dfrac{\alpha \hbar}{c}}$ = 1.602,176,564,374,559,619,029,512,522,979,9e-19 C

$\dfrac{1}{e} = \sqrt{\dfrac{c}{\alpha \hbar}}$ = 6.241,509,345,696,672,315,890,093,396,596,6e18 Coulomb

e^2 = 2.566,969,743,431,067,383,000,001,685,533,2e-38 Coulomb

$= \dfrac{\alpha \hbar}{c} = \dfrac{2h\alpha}{c\mu_0} = \alpha \hbar \sqrt{\varepsilon_0 \mu_0} = \alpha \hbar c \varepsilon_0 \mu_0$

$\dfrac{1}{e^2}$ = 3.895,643,891,241,890,256,554,697,217,129,1e44 Coulomb

$= \dfrac{c}{\alpha \hbar} = \dfrac{c\mu_0}{2h\alpha} = \dfrac{1}{\alpha \hbar \sqrt{\varepsilon_0 \mu_0}} = \dfrac{1}{\alpha \hbar c \varepsilon_0 \mu_0}$

$e^4 = \dfrac{\alpha^2 \hbar^2}{c^2} = \dfrac{e^2 \alpha \hbar}{c} = \alpha^2 \hbar^2 \varepsilon_0 \mu_0$ = 6.589,333,663,690,559,907,896,212,880,112e-90 C

$\dfrac{1}{e^4} = \dfrac{c^2}{\alpha^2 \hbar^2} = \dfrac{c}{e^2 \alpha \hbar} = \dfrac{1}{\alpha^2 \hbar^2 \varepsilon_0 \mu_0}$ = 1.517,604,132,737,025,648,153,796,873,874,3e89 C

$e^8 = \dfrac{e^4 \alpha^2 \hbar^2}{c^2} = \alpha^4 \hbar^4 (\varepsilon_0 \mu_0)^2$ = 4.341,931,813,144,565,686,431,426,306,866,6e-179 C

$e^{12} = \dfrac{e^6 \alpha^3 \hbar^3}{c^3}$ = 2.861,043,746,180,247,659,609,404,757,843,4e-226 C

$e^{16} = \dfrac{e^8 \alpha^4 \hbar^4}{c^4}$ = 1.885,237,186,999,685,567,499,048,388,72e-301 C

Fine Structure Constant α:

$\alpha = \dfrac{c e^2}{\hbar} = \dfrac{e^2}{\hbar c \varepsilon_0 \mu_0}$ = 7.297,352,569,800,184,997,640,004,791,553,6e-3

$= \dfrac{c\mu_0}{2R_K} = \dfrac{ce^2 \mu_0}{2h} = \dfrac{Z_0 G_0}{4} = \dfrac{e^2}{\hbar \sqrt{\varepsilon_0 \mu_0}}$

$$\frac{1}{\alpha} = \frac{\hbar}{c e^2} = \frac{\hbar c \varepsilon_0 \mu_0}{e^2} = 137.035,999,074,302,894,546,650,105,748,4$$

$$= \frac{2R_K}{c\mu_0} = \frac{2h}{ce^2\mu_0} = \frac{4}{Z_0 G_0} = \frac{\hbar\sqrt{\varepsilon_0\mu_0}}{e^2}$$

$$\alpha^2 = \frac{c^2 e^4}{\hbar^2} = \frac{\alpha c e^2}{\hbar} = \frac{e^4}{\hbar^2 \varepsilon_0 \mu_0} = 5.325,135,452,796,936,385,804,739,183,608\text{e-}5$$

$$\frac{1}{\alpha^2} = \frac{\hbar^2}{c^2 e^4} = \frac{\hbar}{\alpha c e^2} = \frac{\hbar^2 \varepsilon_0 \mu_0}{e^4} = 1.877,866,504,229,234,377,110,461,882,738,6$$

$$\alpha^4 = \frac{c^4 e^8}{\hbar^4} = \frac{\alpha^2 c^2 e^4}{\hbar^2} = \frac{e^8}{\hbar^4 (\varepsilon_0\mu_0)^2} = 2.835,706,759,063,483,270,671,023,995,550,9\text{e-}9$$

$$\sqrt{\alpha} = 2.701,361,243,854,695,009,761,325,214,926,6$$

Planck Constant h:

The Planck constant contains the numerical constants "2" and "π" which provide the quantum basis for our decimal system and for our space-time geometry.

$$h = 2\pi\hbar = 6.626,069,570,023,165,074,017,466,569,367,9\text{e-}34 \text{ J·S}$$

$$= \frac{\sqrt{\pi\varepsilon_0}\hbar}{c\varepsilon_0} = \frac{\pi\hbar}{\sqrt{\pi\varepsilon_0}c} = \frac{\sqrt{\pi\varepsilon_0}e^2}{\alpha\varepsilon_0} = \frac{\pi e^2}{\sqrt{\pi\varepsilon_0}\alpha} = \frac{e^2}{2\alpha c\varepsilon_0} = \frac{2\pi c e^2}{\alpha} = \frac{2\pi e^2}{\alpha\sqrt{\varepsilon_0\mu_0}} = \frac{2\pi e^2}{\alpha c\varepsilon_0\mu_0}$$

$$h^2 = \frac{\pi\hbar^2}{c^2\varepsilon_0} = \frac{\pi e^4}{\alpha^2\varepsilon_0} = 4.390,479,794,678,697,168,406,840,238,821,6\text{e-}17 \text{ J·S}$$

$$h^3 = \frac{\sqrt{\pi\varepsilon_0}\pi\hbar^3}{c^3\varepsilon_0^2} = = 2.909,162,456,532,206,902,424,165,293,232,9$$

$$= \frac{\pi^2\hbar^3}{\sqrt{\pi\varepsilon_0}c^3\varepsilon_0} = \frac{\sqrt{\pi\varepsilon_0}\pi\hbar^2 e^2}{\alpha c^2\varepsilon_0^2} = \frac{\pi^2\hbar^2 e^2}{\sqrt{\pi\varepsilon_0}\alpha c^2\varepsilon_0} = \frac{\sqrt{\pi\varepsilon_0}\pi e^4\hbar}{\alpha^2 c\varepsilon_0^2} = \frac{\pi^2 e^4\hbar}{\sqrt{\pi\varepsilon_0}\alpha^2 c\varepsilon_0} = \frac{\sqrt{\pi\varepsilon_0}\pi e^6}{\alpha^3\varepsilon_0^2} = \frac{\pi^2 e^6}{\sqrt{\pi\varepsilon_0}\alpha^3\varepsilon_0}$$

$$2h = \frac{\hbar}{c^2\varepsilon_0} = \frac{e^2}{\alpha c\varepsilon_0} = 1.325,213,914,004,633,014,803,493,313,873,6\text{e-}33 \text{ J·S}$$

$$= 4\pi\hbar = \hbar\mu_0 = \frac{4\pi e^2 c}{\alpha} = \frac{4\pi e^2}{\alpha\sqrt{\varepsilon_0\mu_0}}$$

Reduced Planck Constant $\hbar$:

$$\hbar = \frac{c\,e^2}{\alpha} = 1.054{,}571{,}725{,}339{,}976{,}234{,}000{,}102{,}692{,}456{,}8\text{e-}34 \text{ J·S}$$

$$= \frac{2h}{\mu_0} = \frac{e^2}{\alpha\sqrt{\varepsilon_0\mu_0}} = \frac{e^2}{\alpha c \varepsilon_0 \mu_0}$$

$$\frac{1}{\hbar} = \frac{\alpha}{c\,e^2} = 9.482{,}522{,}392{,}468{,}058{,}773{,}514{,}969{,}434{,}031{,}2\text{e}33 \text{ J·S}$$

$$= \frac{\mu_0}{2h} = \frac{\alpha\sqrt{\varepsilon_0\mu_0}}{e^2} = \frac{\alpha c \varepsilon_0 \mu_0}{e^2}$$

$$\hbar^2 = \frac{c^2 e^4}{\alpha^2} = \frac{\hbar c e^2}{\alpha} = \frac{e^4}{\alpha^2 \varepsilon_0 \mu_0} = 1.112{,}121{,}523{,}886{,}534{,}272{,}212{,}567{,}732{,}807{,}5\text{e-}68 \text{ J·S}$$

$$\frac{1}{\hbar^2} = \frac{\alpha^2}{c^2 e^4} = \frac{\alpha}{\hbar c e^2} = \frac{\alpha^2 \varepsilon_0 \mu_0}{e^4} = 8.991{,}823{,}092{,}365{,}815{,}726{,}547{,}458{,}742{,}625{,}8\text{e}67 \text{ J·S}$$

$$\hbar^4 = \frac{c^4 e^8}{\alpha^4} = \frac{\hbar^2 c^2 e^4}{\alpha^2} = \frac{e^8}{\alpha^4 (\varepsilon_0 \mu_0)^2} = .123{,}681{,}428{,}389{,}170{,}721{,}979{,}541{,}782{,}930{,}85 \text{ J·S}$$

Numerical constants:

$$\pi = \frac{h}{2\hbar} = 3.141{,}592{,}653{,}589{,}793{,}238{,}462{,}643{,}383{,}279{,}5$$

$$\pi = 4\pi^2 c^2 \varepsilon_0 = \frac{e^4}{4\alpha^2 \hbar^2 \varepsilon_0} = \frac{1}{4c^2 \varepsilon_0} = \frac{\alpha h}{2c e^2} = \frac{c e^2}{2\alpha \hbar}$$

$$1 = \frac{\pi c \varepsilon_0}{(\sqrt{\pi \varepsilon_0} - \pi c \varepsilon_0)} = c^2 \varepsilon_0 \mu_0 = \frac{\alpha \hbar}{c e^2} = \frac{c e^2}{\alpha \hbar} = \frac{e^4}{\alpha^2 \hbar^2 \varepsilon_0 \mu_0} = \frac{\alpha^2 \hbar^2 \varepsilon_0 \mu_0}{e^4}$$

$$2 = \frac{1}{\sqrt{\pi \varepsilon_0}\, c} = \frac{\alpha^2 \hbar^2}{\sqrt{\pi \varepsilon_0}\, e^4 c^3}$$

$$3 = (\sqrt{\pi \varepsilon_0} + \pi c \varepsilon_0)\, c$$

$$4 = \frac{1}{\pi c^2 \varepsilon_0} = \frac{e^4}{\pi \alpha^2 \hbar^2 \varepsilon_0} = \frac{(\pi c \varepsilon_0 - \sqrt{\pi \varepsilon_0})}{\sqrt{\pi \varepsilon_0}} = \frac{1}{\pi c^4 \varepsilon_0^2 \mu_0}$$

$$5 = \sqrt{\pi \varepsilon_0}\, c = \frac{\sqrt{\pi \varepsilon_0}\, e^4 c^3}{\alpha^2 \hbar^2} = \frac{\sqrt{\pi \varepsilon_0}\, c^2 e^2}{\alpha \hbar} = \frac{\sqrt{\pi \varepsilon_0}\, \alpha \hbar}{e^2} = \frac{\sqrt{\pi \varepsilon_0}}{c\, \varepsilon_0 \mu_0}$$

$$6 = \frac{(\sqrt{\pi \varepsilon_0} + \pi c \varepsilon_0)}{\sqrt{\pi \varepsilon_0}}$$

$$7 = \frac{1}{\sqrt{\pi \varepsilon_0}\, c} + \sqrt{\pi \varepsilon_0}\, c$$

$$8 = \frac{1}{\sqrt{\pi \varepsilon_0}\, \pi c^3 \varepsilon_0} = \frac{(\pi c \varepsilon_0 - \sqrt{\pi \varepsilon_0})}{\pi c \varepsilon_0} = \frac{(\pi c \varepsilon_0 - \sqrt{\pi \varepsilon_0})}{(\sqrt{\pi \varepsilon_0} - \pi c \varepsilon_0)}$$

$$9 = (\sqrt{\pi \varepsilon_0} + \pi c \varepsilon_0)^2 c^2$$

$$10 = \frac{\sqrt{\pi \varepsilon_0}\, c}{\sqrt{\pi \varepsilon_0}\, c}$$

$$11 = \frac{(\sqrt{\pi \varepsilon_0} + \pi c \varepsilon_0)}{\sqrt{\pi \varepsilon_0}} + \sqrt{\pi \varepsilon_0}\, c$$

$$12 = \frac{(\sqrt{\pi \varepsilon_0} + \pi c \varepsilon_0)}{\pi c \varepsilon_0} = \frac{(\sqrt{\pi \varepsilon_0} + \pi c \varepsilon_0)}{(\sqrt{\pi \varepsilon_0} - \pi c \varepsilon_0)}$$

$$13 = \frac{\sqrt{\pi \varepsilon_0}}{\pi^2 c^3 \varepsilon_0^2} + \sqrt{\pi \varepsilon_0}\, c$$

$$14 = \left(\frac{1}{\sqrt{\pi \varepsilon_0}\, c} + \sqrt{\pi \varepsilon_0}\, c\right) \frac{1}{\sqrt{\pi \varepsilon_0}\, c}$$

$$15 = (\sqrt{\pi \varepsilon_0} + \pi c \varepsilon_0)\sqrt{\pi \varepsilon_0}\, c^2$$

$$16 = \frac{1}{\pi^2 c^4 \varepsilon_0^2}$$

$$18 = \frac{(\sqrt{\pi \varepsilon_0} + \pi c \varepsilon_0)^2 c}{\sqrt{\pi \varepsilon_0}}$$

$$20 = \frac{\sqrt{\pi \varepsilon_0}}{\pi c \varepsilon_0}$$

$$24 = \frac{(\sqrt{\pi\varepsilon_0} + \pi c\varepsilon_0)}{\sqrt{\pi\varepsilon_0}\,\pi c^2\varepsilon_0} = \frac{(\sqrt{\pi\varepsilon_0} + \pi c\varepsilon_0)(\pi c\varepsilon_0 - \sqrt{\pi\varepsilon_0})}{\pi\varepsilon_0}$$

$$25 = \pi c^2\varepsilon_0 = \frac{\pi\alpha^2\hbar^2\varepsilon_0}{e^4} = \frac{\sqrt{\pi\varepsilon_0}}{(\pi c\varepsilon_0 - \sqrt{\pi\varepsilon_0})} = \pi c^4\varepsilon_0^2\mu_0$$

$$27 = (\sqrt{\pi\varepsilon_0} + \pi c\varepsilon_0)^3 c^3$$

$$32 = \frac{1}{\sqrt{\pi\varepsilon_0}\,\pi^2 c^5\varepsilon_0^2}$$

$$36 = \frac{(\sqrt{\pi\varepsilon_0} + \pi c\varepsilon_0)^2}{\pi\varepsilon_0}$$

$$48 = \frac{(\sqrt{\pi\varepsilon_0} + \pi c\varepsilon_0)}{\pi^2 c^3\varepsilon_0^2}$$

$$50 = \frac{\pi c\varepsilon_0}{\sqrt{\pi\varepsilon_0}}$$

$$64 = \frac{1}{\pi^3 c^6\varepsilon_0^3}$$

$$72 = \frac{(\sqrt{\pi\varepsilon_0} + \pi c\varepsilon_0)^2\sqrt{\pi\varepsilon_0}}{\pi^2 c\varepsilon_0^2}$$

$$75 = (\sqrt{\pi\varepsilon_0} + \pi c\varepsilon_0)\pi c^3\varepsilon_0 = \frac{(\sqrt{\pi\varepsilon_0} + \pi c\varepsilon_0)\pi e^4 c^5\varepsilon_0}{\alpha^2\hbar^2}$$

$$80 = \frac{\sqrt{\pi\varepsilon_0}}{\pi^2 c^3\varepsilon_0^2}$$

$$81 = (\sqrt{\pi\varepsilon_0} + \pi c\varepsilon_0)^4 c^4$$

$$125 = \sqrt{\pi\varepsilon_0}\,\pi c^3\varepsilon_0$$

$$128 = \frac{\sqrt{\pi\varepsilon_0}}{\pi^4 c^7\varepsilon_0^4}$$

$$144 = \frac{(\sqrt{\pi\varepsilon_0} + \pi c\varepsilon_0)^2}{\pi^2 c^2\varepsilon_0^2}$$

$$243 = (\sqrt{\pi\varepsilon_0} + \pi c\varepsilon_0)^5 c^5$$

$$256 = \frac{1}{\pi^4 c^8 \varepsilon_0^4}$$

$$375 = (\sqrt{\pi\varepsilon_0} + \pi c\varepsilon_0)\sqrt{\pi\varepsilon_0}\,\pi c^4 \varepsilon_0$$

$$625 = \pi^2 c^4 \varepsilon_0^2$$

$$1{,}024 = \frac{1}{\pi^5 c^{10} \varepsilon_0^5}$$

$$3{,}125 = \sqrt{\pi\varepsilon_0}\,\pi^2 c^5 \varepsilon_0^2$$

$$15{,}625 = \pi^3 c^6 \varepsilon_0^3$$

$$78{,}125 = \frac{\pi^4 c^7 \varepsilon_0^4}{\sqrt{\pi\varepsilon_0}}$$

$$390{,}625 = \pi^4 c^8 \varepsilon_0^4$$

Coulomb constant: $k_e = c^2 = \dfrac{\alpha^2 \hbar^2}{e^4} = \dfrac{1}{\varepsilon_0 \mu_0} = 8.987{,}551{,}787{,}368{,}176{,}4\text{E}9 \ \text{N·m}^2/\text{C}^2$

Electromagnetic coupling constant:

$$g = \frac{e}{\sqrt{\hbar c}} = \frac{\sqrt{\alpha}}{c} = \sqrt{\alpha\,\varepsilon_0 \mu_0} = 2.849{,}456{,}043{,}849{,}631{,}373{,}051{,}310{,}841{,}799$$

$$g^2 = \frac{e^2}{\hbar c} = \frac{\alpha}{c^2} = \alpha\,\varepsilon_0 \mu_0 = 8.119{,}399{,}745{,}831{,}192{,}350{,}248{,}763{,}524{,}544$$

Von Klitzing constant:

$$R_K = \frac{h}{e^2} = \frac{1}{2\alpha c \varepsilon_0} = 25{,}812.807{,}443{,}404{,}521{,}417{,}322{,}882{,}144{,}935 \text{ Ohms}$$

$$R_K = \frac{\pi \hbar}{\sqrt{\pi \varepsilon_0} e^2 c} = \frac{\sqrt{\pi \varepsilon_0}}{\alpha \varepsilon_0} = \frac{\pi}{\sqrt{\pi \varepsilon_0} \alpha} = \frac{c \mu_0}{2\alpha} = \frac{e^2 \mu_0}{8\pi \alpha^2 \hbar \varepsilon_0} = \frac{2\pi}{\alpha \sqrt{\varepsilon_0 \mu_0}} = \frac{2\pi}{\alpha c \varepsilon_0 \mu_0}$$

Hall effect: $= \dfrac{e^2}{h} = 2\alpha c \varepsilon_0 = .387{,}404{,}586{,}731{,}813{,}189{,}681{,}564{,}599{,}955{,}21$

$$= \frac{\sqrt{\pi \varepsilon_0} e^2 c}{\pi \hbar} = \frac{\alpha \varepsilon_0}{\sqrt{\pi \varepsilon_0}} = \frac{\sqrt{\pi \varepsilon_0} \alpha}{\pi} = \frac{2\alpha}{c\mu_0} = \frac{8\pi \alpha^2 \hbar \varepsilon_0}{e^2 \mu_0} = \frac{\alpha \sqrt{\varepsilon_0 \mu_0}}{2\pi} = \frac{\alpha c \varepsilon_0 \mu_0}{2\pi}$$

Josephson constant:

$$K_J = \frac{2e}{h} = \frac{e}{\pi \hbar} = \frac{2\sqrt{\alpha \hbar}}{h\sqrt{c}} = 483{,}597.869{,}730{,}473{,}816{,}563{,}327{,}973{,}661{,}2 \text{ Hz/Volt}$$

$$\Phi_0 = \frac{h}{2e} = \frac{\pi \hbar}{e} = \frac{h\sqrt{c}}{2\sqrt{\alpha \hbar}} = 2.067{,}833{,}757{,}326{,}795{,}795{,}427{,}727{,}403{,}567{,}2$$

Gravitational constant:

G $=$ $6.672{,}536{,}659{,}401{,}633{,}510{,}465{,}095{,}670{,}374{,}1\text{e-}11 \text{ m}^3 \text{Kg}^{-1} \text{sec}^{-2}$

$$G = \frac{\hbar c}{8\pi e^{16}} = \frac{\pi \hbar c^4 \varepsilon_0^2}{\sqrt{\pi \varepsilon_0}\, e^{16}} = \frac{\sqrt{\pi \varepsilon_0}\, \hbar c^4 \varepsilon_0}{e^{16}} = \frac{c^5}{8\pi e^8 \alpha^4 \hbar^3} = \frac{\pi c^8 \varepsilon_0^2}{\sqrt{\pi \varepsilon_0}\, e^8 \alpha^4 \hbar^3} = \frac{\sqrt{\pi \varepsilon_0}\, c^8 \varepsilon_0}{e^8 \alpha^4 \hbar^3}$$

G^2 $=$ $4.452{,}274{,}547{,}105{,}871{,}092{,}528{,}373{,}147{,}775{,}5\text{e-}21 \text{ m}^3 \text{Kg}^{-1} \text{sec}^{-2}$

$$G^2 = \frac{\hbar^2 c^2}{64 \pi^2 e^{32}} = \frac{c^{10}}{64 \pi^2 e^{16} \alpha^8 \hbar^6}$$

Gm_e^2 $=$ $=$ $5.536{,}928{,}095{,}587{,}838{,}387{,}937{,}489{,}950{,}241{,}4$

Gravitational coupling constant:

$$\alpha_G = \frac{Gm_e^2}{\hbar c} = \qquad = 1.751,346,414,855,346,417,352,362,154,042,5\text{e-}45$$

Boltzmann constant:

$$K_B = R\, m_u = \frac{R}{N_A} = \frac{R\, m_e}{A_r(e)} = 1.380,648,798,394,132,924,292,774,930,237,3\text{e-}23 \text{ J/K}$$

$$K_B^2 = \qquad = 1.906,191,104,507,163,100,509,430,177,044,9\text{e-}42 \text{ J/K}$$

$$K_B^4 = \frac{R^4}{N_A^4} = \qquad = 3.633,564,526,902,238,397,193,481,753,702,8\text{e-}81 \text{ J/K}$$

Stefan-Boltzmann constant:

$$\sigma = \frac{\sqrt{\pi\varepsilon_0}\,\pi^2 R^4}{(\sqrt{\pi\varepsilon_0} + \pi c\varepsilon_0)\hbar^3 c^2 N_A^4} = 5.670,372,596,150,452,860,287,891,723,617\text{e-}8 \text{ W}_m$$

$$= \frac{\sqrt{\pi\varepsilon_0}\,\pi^2 K_B^4}{(\sqrt{\pi\varepsilon_0} + \pi c\varepsilon_0)\hbar^3 c^2} = \frac{\pi^2 K_B^4}{60\hbar^3 c^2} = \frac{2\pi^5 K_B^4}{15 c^2 h^3} = \frac{2\pi^5 K_B^4}{(\sqrt{\pi\varepsilon_0}+\pi c\varepsilon_0)\sqrt{\pi\varepsilon_0}\,h^3 c^4} =$$

$$\frac{\pi^4 K_B^4}{(\sqrt{\pi\varepsilon_0}+\pi c\varepsilon_0)h^3 c^5 \varepsilon_0} = \frac{\pi^3 \alpha \varepsilon_0 K_B^4}{(\sqrt{\pi\varepsilon_0}+\pi c\varepsilon_0)\sqrt{\pi\varepsilon_0}\,e^2 \hbar^2 c^3} = \frac{\sqrt{\pi\varepsilon_0}\,\pi^2 \alpha K_B^4}{(\sqrt{\pi\varepsilon_0}+\pi c\varepsilon_0)e^2 \hbar^2 c^3} =$$

$$\frac{\pi^3 \alpha^2 \varepsilon_0 K_B^4}{(\sqrt{\pi\varepsilon_0}+\pi c\varepsilon_0)\sqrt{\pi\varepsilon_0}\,e^4 \hbar c^4} = \frac{\sqrt{\pi\varepsilon_0}\,\pi^2 \alpha^2 K_B^4}{(\sqrt{\pi\varepsilon_0}+\pi c\varepsilon_0)e^4 \hbar c^4} = \frac{\pi^3 \alpha^3 \varepsilon_0 K_B^4}{(\sqrt{\pi\varepsilon_0}+\pi c\varepsilon_0)\sqrt{\pi\varepsilon_0}\,e^6 c^5} =$$

$$\frac{\sqrt{\pi\varepsilon_0}\,\pi^2 \alpha^3 K_B^4}{(\sqrt{\pi\varepsilon_0}+\pi c\varepsilon_0)e^6 c^5} = \frac{\pi^3 \varepsilon_0 K_B^4}{(\sqrt{\pi\varepsilon_0}+\pi c\varepsilon_0)\sqrt{\pi\varepsilon_0}\,\hbar^3 c^2} = \frac{32\pi^5 h R_\infty^4 R^4}{15 A_r(e)^4 M_u^4 \alpha^8 c^6} =$$

$$\frac{\pi^3 \hbar R_\infty^4 R^4}{(\sqrt{\pi\varepsilon_0}+\pi c\varepsilon_0)\sqrt{\pi\varepsilon_0}\,A_r(e)^4 M_u^4 c^8 \varepsilon_0^3} = \frac{\pi^3 e^2 R_\infty^4 R^4}{(\sqrt{\pi\varepsilon_0}+\pi c\varepsilon_0)\sqrt{\pi\varepsilon_0}\,A_r(e)^4 M_u^4 \alpha c^7 \varepsilon_0^3}$$

Mass of an electron:

$$m_e = \frac{e^2}{r_e} = \frac{h}{\lambda_e c} = 9.109,382,903,345,524,937,498,006,139,342,6\text{e-}31 \text{ Kg}$$

$$= \frac{\pi\hbar}{\lambda_e\sqrt{\pi\varepsilon_0}\,c^2} = \frac{\pi e^2}{\lambda_e\sqrt{\pi\varepsilon_0}\,\alpha c} = \frac{\sqrt{\pi\varepsilon_0}\,e^2}{\lambda_e\alpha c\varepsilon_0} = \frac{R_\infty \hbar}{\alpha^2 c^3 \varepsilon_0} = \frac{R_\infty \mu_0 \hbar}{\alpha^2 c} = \frac{R_\infty 4\pi\hbar\sqrt{\varepsilon_0\mu_0}}{\alpha^2}$$

$$m_e^2 = 8.298,085,687,976,374,532,552,834,892,706,6\text{e-}61 \text{ Kg}$$

$$m_e^2 = \frac{h^2}{\lambda_e^2 c^2} = \frac{4\pi^2 e^4}{\alpha^2 \lambda_e^2} = \frac{\pi\hbar^2}{\lambda_e^2 c^4 \varepsilon_0} = \frac{\pi e^4}{\lambda_e^2 \alpha^2 c^2 \varepsilon_0} = \frac{\pi e^2}{\lambda_e^2 \alpha c}$$

Energy equivalent for an electron:

$$m_e c^2 = \quad = 8.187,105,059,478,418,073,403,857,171,418,8\text{e-}14 \text{ J}$$

Compton wavelength of an electron:

$$\lambda_e = \frac{h}{m_e c} = 2.426,310,238,899,999,814,307,976,139,616,8\text{e-}12 \text{ meter}$$

$$\lambda_e = \frac{\pi\hbar}{m_e\sqrt{\pi\varepsilon_0}\,c^2} = \frac{\pi e^2}{m_e\sqrt{\pi\varepsilon_0}\,\alpha c} = \frac{\sqrt{\pi\varepsilon_0}\,e^2}{m_e \alpha c \varepsilon_0}$$

$$\lambda_e c = \quad = 7.273,895,103,903,981,605,309,317,359,010,7$$

$$\lambda_e^2 = \frac{h^2}{m_e^2 c^2} = 5.886,981,375,390,974,172,117,082,450,978,3\text{e-}24 \text{ meter}$$

$$\lambda_e^2 = \frac{\hbar^2}{m_e^2 c^4 \varepsilon_0} = \frac{e^2}{m_e^2 \alpha c \varepsilon_0} = \frac{\pi e^2}{m_e^2 \alpha c}$$

Compton frequency (electron):

$$f_e = \frac{m_e c^2}{h} = \frac{m_e}{h\varepsilon_0 \mu_0} = 1.235,589,963,696,954,594,522,855,030,459,7e20 \text{ sec}^{-1}$$

$$f_e = \frac{\sqrt{\pi\varepsilon_0} c^3 m_e}{\pi \hbar} = \frac{\alpha c^2 \varepsilon_0 m_e}{\sqrt{\pi\varepsilon_0}\, e^2} = \frac{\sqrt{\pi\varepsilon_0}\, \alpha c^2 m_e}{\pi e^2}$$

$$f_e^2 = \frac{m_e^2 c^4}{h^2} = \frac{m_e^2 c^6 \varepsilon_0}{\pi \hbar^2} = 1.526,682,558,388,641,572,804,015,098,630,5e40 \text{ sec}^{-1}$$

$$f_e^2 = \frac{m_e^2 \alpha^2 c^4 \varepsilon_0}{\pi e^4}$$

Bohr radius (electron orbit):

$$a_0 = 5.291,772,109,155,282,349,118,932,383,177,9e-11 \text{ m}$$

$$a_0 = \frac{\lambda_e}{2\pi\alpha} = \frac{\lambda_e c \varepsilon_0}{\sqrt{\pi\varepsilon_0}\,\alpha} = \frac{\lambda_e \sqrt{\pi\varepsilon_0}\, c}{\pi\alpha} = \frac{2\lambda_e \hbar \varepsilon_0}{e^2 \sqrt{\varepsilon_0 \mu_0}}$$

$$a_0^2 = 2.800,285,205,523,374,548,919,924,682,626,4e-21 \text{ m}$$

$$a_0^2 = \frac{\lambda_e^2}{2\pi^2 \alpha^2}$$

Compton angular frequency (electron):

$$\omega_e = 1.552,688,141,119,852,751,491,344,890,587,9e21 \text{ radians/sec}$$

$$\omega_e = \frac{2 m_e c^2}{\hbar} = \frac{m_e c}{\sqrt{\pi\varepsilon_0}\,\hbar} = \frac{m_e \sqrt{\pi\varepsilon_0}\, c}{\pi \hbar \varepsilon_0} = \frac{2 m_e}{\hbar \varepsilon_0 \mu_0} = \frac{m_e}{\sqrt{\pi\varepsilon_0}\, \hbar c \varepsilon_0 \mu_0} = \frac{m_e \sqrt{\pi\varepsilon_0}}{\pi c \varepsilon_0^2 \mu_0}$$

$$\omega_e^2 = 2.410,840,463,574,223,772,828,027,233,553,6e42 \text{ radians/sec}$$

$$\omega_e^2 = \frac{4 m_e^2 c^4}{\hbar^2} = \frac{m_e^2 c^2}{\pi \hbar^2 \varepsilon_0}$$

Classical electron (orbit) radius:

r_e = 2.817,940,326,658,481,354,953,560,028,397,4e-15 meter

$$r_e = \frac{e^2}{m_e} = \frac{\alpha \lambda_e}{2\pi} = \frac{\alpha c \varepsilon_0 \lambda_e}{\sqrt{\pi \varepsilon_0}} = \frac{\sqrt{\pi \varepsilon_0}\, \alpha c \lambda_e}{\pi}$$

r_e^2 = 7.940,787,684,608,108,604,519,111,017,387e-30 meter

$$r_e^2 = \frac{e^4}{m_e^2} = \frac{\alpha^2 \lambda_e^2}{4\pi^2}$$

Relative mass of an electron:

$A_r(e)$ = 5.485,799,109,910,795,715,106,472,125,503,7e-4 Kg

$$A_r(e) = \frac{m_e}{m_u} = \frac{m_e c \alpha^2}{m_u c \alpha^2}$$

$$A_r(e)^2 = \frac{m_e^2}{m_u^2} = 3.009,399,187,429,807,852,665,375,408,709,9\text{e-7 Kg}$$

$A_r(e)^4$ = = 9.056,483,469,303,187,773,939,527,866,255,8e-14 Kg

Electron frequency:

$$m_e c \alpha^2 = 2hR_\infty = \frac{\hbar R_\infty}{c^2 \varepsilon_0} = 1.454,254,176,334,272,379,456,085,011,440,9\text{e20 Hz}$$

$$= 4\pi \hbar R_\infty = \frac{2\pi e^2 c \alpha}{\lambda_e} = \frac{\pi e^2 \alpha}{\sqrt{\pi \varepsilon_0}\, \lambda_e} = \frac{\sqrt{\pi \varepsilon_0}\, e^2 \alpha}{\varepsilon_0 \lambda_e}$$

Nucleon frequency:

$m_u c \alpha^2$ = 2.650,943,184,753,114,170,389,320,546,962,5e23 Hz

$$m_u c \alpha^2 = \frac{m_e c \alpha^2}{A_r(e)} = \frac{2hR_\infty}{A_r(e)} = \frac{\hbar R_\infty}{c^2 \varepsilon_0 A_r(e)} = \frac{e^2 \alpha^2 c}{r_e A_r(e)} = \frac{\alpha^3 \hbar}{r_e A_r(e)}$$

Frequency difference between an electron and a nucleon (proton/neutron):

$$\frac{m_u c\alpha^2}{m_e c\alpha^2} = \frac{m_u}{m_e} = \frac{1}{A_r(e)} = 1.822,888,479,808,479,436,514,029,682,372 \text{ Hz}$$

$$= \frac{m_u c\alpha^2}{2hR_\infty} = \frac{m_u \alpha^2 c^3 \varepsilon_0}{\hbar R_\infty} = \frac{\sqrt{\pi\varepsilon_0}\alpha c\lambda_e m_u}{\pi e^2} = \frac{\alpha c\varepsilon_0 \lambda_e m_u}{\sqrt{\pi\varepsilon_0} e^2} = \frac{r_e m_u c}{\alpha \hbar} = \frac{r_e m_u}{e^2}$$

Atomic mass unit (mass of a nucleon):

$$m_u = 1.660,538,915,267,287,672,145,963,619,188,4\text{e-27 Kg}$$

$$m_u = \frac{1}{N_A} = \frac{4\pi\hbar R_\infty}{A_r(e)c\alpha^2} = \frac{m_e}{A_r(e)} = \frac{\hbar R_\infty}{c^2 \varepsilon_0 A_r(e)}$$

$$m_u^2 = \frac{m_e^2}{A_r(e)^2} = 2.757,354,423,471,677,403,524,097,448,165,7\text{e-54 Kg}$$

Energy equivalent for a nucleon:

$$m_u c^2 = \quad = 1.492,417,949,590,492,414,032,154,046,336,9\text{e-10 J}$$

Compton wavelength (nucleon):

$$\lambda_p = \frac{h}{m_u c} = A_r(e)\lambda_e = 1.331,025,054,892,506,909,053,994,151,094,3\text{e-15m}$$

Rydberg constant: $R_\infty = 1.097,373,156,866,196,496,466,432,160,158,9\text{E7/m}$

$$R_\infty = \frac{\alpha^2}{2\lambda_e} = \frac{m_e c\alpha^2}{\mu_0 \hbar} = \frac{m_e \alpha^2}{4\pi\hbar\sqrt{\varepsilon_0 \mu_0}} = \frac{m_e e^4}{8\varepsilon_0^2 h^3 c}$$

$R_\infty^2 = 1.204,227,845,410,481,902,837,319,385,797,7\text{E14/m}$

Gas Constant: R = 8.314,462,164,663,558,038,800,395,847,263,8 J/Mol·K

$$R = K_B N_A = \frac{K_B}{m_u} = \frac{A_r(e) K_B}{m_e}$$

Avogadro number:

$$N_A = 602,214,131,090,710,142,596,905 \text{ Mol}^{-1}$$

$$N_A = \frac{A_r(e)}{m_e} = \frac{1}{m_u} = \frac{m_u A_r(e) c \alpha^2}{2 h R_\infty}$$

$$N_A^4 = 13,152,362,447,042,852,783,813,938,417,055 \text{ Mol}^{-1}$$

Planck mass:

$$m_P = 2.176,721,809,869,602,870,921,689,401,314,3\text{e-}8 \text{ Kg}$$

$$m_P = \sqrt{\frac{\hbar c}{G}} = \sqrt{8\pi}\, e^8 = \frac{\sqrt{8\pi}\, e^4 \alpha^2 \hbar^2}{c^2}$$

$$m_P^2 = 4.738,117,837,561,999,550,364,715,655,045,2\text{e-}16 \text{ Kg}$$

$$m_P^2 = \frac{\hbar c}{G} = 8\pi e^{16} = \frac{\sqrt{\pi \varepsilon_0}\, e^{16}}{\pi c^3 \varepsilon_0^2} = \frac{e^{16}}{\sqrt{\pi \varepsilon_0}\, c^3 \varepsilon_0} = \frac{8\pi e^8 \alpha^4 \hbar^4}{c^4} = \frac{\sqrt{\pi \varepsilon_0}\, e^8 \alpha^4 \hbar^4}{\pi c^7 \varepsilon_0^2} = \frac{e^8 \alpha^4 \hbar^4}{\sqrt{\pi \varepsilon_0}\, c^7 \varepsilon_0}$$

Planck length:

$$l_p = 1.616,041,433,451,048,138,097,547,554,248,1\text{e-}35 \text{ meter}$$

$$l_p = c\, t_P = \sqrt{\frac{\hbar G}{c^3}} = \frac{\lambda_e \sqrt{\alpha_G}}{2\pi} = \frac{c}{\sqrt{8\pi}\, e^4 \alpha^2 \hbar}$$

Planck area:
$$l_p^2 = 2.611,589,914,630,518,448,089,729,709,586,8\text{e-}70 \text{ meter}^2$$

$$l_p^2 = \frac{\hbar G}{c^3} = \frac{\lambda_e^2 \alpha_G}{4\pi^2} = \frac{c^2}{8\pi e^8 \alpha^4 \hbar^2} = \frac{\pi c^5 \varepsilon_0^2}{\sqrt{\pi \varepsilon_0}\, e^8 \alpha^4 \hbar^2} = \frac{\sqrt{\pi \varepsilon_0}\, c^5 \varepsilon_0}{e^8 \alpha^4 \hbar^2}$$

Planck volume:

l_p^3 = 4.220,437,509,225,803,466,852,291,425,321,7e-105 meter³

$$l_p^3 = \sqrt{\frac{(\hbar G)^3}{c^9}} = \frac{c^5\varepsilon_0}{2\sqrt{8\pi}\, e^{12}\alpha^6\hbar^3} = \frac{\sqrt{\pi\varepsilon_0}\,c^6\varepsilon_0}{\sqrt{8\pi}\, e^{12}\alpha^6\hbar^3} = \frac{\pi c^6\varepsilon_0^2}{\sqrt{\pi\varepsilon_0}\sqrt{8\pi}\, e^{12}\alpha^6\hbar^3}$$

Planck time:

t_P = 5.390,533,985,518,235,212,233,216,201,349,9e-44 Sec

$$t_P = \frac{l_p}{c} = \frac{1}{\omega_P} = \frac{\hbar}{m_P c^2} = \sqrt{\frac{\hbar G}{c^5}} = \frac{\hbar}{\sqrt{8\pi}\, e^8 c^2} = \frac{1}{\sqrt{8\pi}\, e^4\alpha^2\hbar}$$

t_P^2 = 2.905,785,664,902,710,927,280,224,709,366,4e-87 Sec

$$t_P^2 = \frac{\hbar^2}{m_P^2 c^4} = \frac{\hbar G}{c^5} = \frac{\sqrt{\pi\varepsilon_0}\,c^3\varepsilon_0}{e^8\alpha^4\hbar^2} = \frac{\pi c^3\varepsilon_0^2}{\sqrt{\pi\varepsilon_0}\,e^8\alpha^4\hbar^2}$$

Planck momentum:

M_P = 6.525,647,817,630,169,041,574,699,911,326,2 $K_g \cdot m/s$

$$M_P = m_P c = \frac{\hbar}{l_p} = \sqrt{\frac{\hbar c^3}{G}} = \sqrt{8\pi}\, e^8 c = \frac{\sqrt{8\pi}\, e^4\alpha^2\hbar^2}{c}$$

Planck energy:

E_P = 1.956,339,999,289,684,111,929,183,477,028,9e9 J

$$E_P = m_P c^2 = \frac{\hbar}{t_P} = \frac{m_P l_p^2}{t_P^2} = K_B T_P = \sqrt{\frac{\hbar c^5}{G}} = \sqrt{8\pi}\, e^4\alpha^2\hbar^2 = \sqrt{8\pi}\, e^8 c^2 = \frac{\sqrt{8\pi}\, e^8}{\varepsilon_0\mu_0}$$

E_P^2 = 3.827,266,192,820,761,231,567,626,267,746,4e18 J

$$E_P^2 = m_P^2 c^4 = \frac{\hbar^2}{t_P^2} = \frac{\hbar c^5}{G} = 8\pi e^{16}c^4 = 8\pi e^8\alpha^4\hbar^4 = \frac{\sqrt{\pi\varepsilon_0}\,e^8\alpha^4\hbar^4}{\pi c^3\varepsilon_0^2} = \frac{e^8\alpha^4\hbar^4}{\sqrt{\pi\varepsilon_0}\,c^3\varepsilon_0}$$

Planck charge:

$$q_P = \sqrt{4\pi\varepsilon_0 \hbar c} = \frac{e}{\sqrt{\alpha}} = 1.875,545,956,207,374,004,255,058,967,960,8\text{e-}18 \text{ C}$$

$$q_P^2 = 4\pi\varepsilon_0 \hbar c = \frac{e^2}{\alpha} = \frac{\hbar}{c} = 3.517,672,633,845,832,886,163,209,257,441,7\text{e-}36 \text{ C}$$

Planck temperature:

$$T_P = 1.416,971,500,330,244,730,973,577,701,559,2\text{e}32 \text{ Kelvin}$$

$$T_P = \frac{m_P c^2}{K_B} = \sqrt{\frac{\hbar c^5}{G K_B^2}}$$

$$T_P^2 = 2.007,808,232,748,144,743,738,515,294,625,1\text{e}64 \text{ Kelvin}$$

$$T_P^2 = \frac{m_P^2 c^4}{K_B^2} = \frac{\hbar c^5}{G K_B^2}$$

Planck density:

$$\rho_P = 5.157,573,841,838,261,099,385,732,637,947\text{E}96 \text{ Kg/m}^3$$

$$\rho_P = \frac{m_P}{l_P^3} = \frac{\hbar t_P}{l_P^5} = \frac{c^5}{\hbar G^2} = \frac{64\pi^2 e^{16} \alpha^8 \hbar^5}{c^5} = \frac{e^{16} \alpha^8 \hbar^5}{\pi c^{11} \varepsilon_0^3}$$

Planck energy density:

$$\rho_P^E = 4.635,396,200,069,681,587,857,900,632,632\text{e}113 \text{ J/m}^3$$

$$\rho_P^E = \frac{E_P}{l_P^3} = \frac{c^7}{\hbar G^2} = \frac{64\pi^2 e^{16} \alpha^8 \hbar^5}{c^3} = \frac{e^{16} \alpha^8 \hbar^5}{\pi c^9 \varepsilon_0^3}$$

Planck intensity:

$$I_P = 1.389,656,820,622,749,614,501,262,985,376,5\text{e}122 \text{ W/m}^2$$

$$I_P = \rho_P^E c = \frac{\rho_P}{l_P^2} = \frac{c^8}{\hbar G^2} = \frac{64\pi^2 e^{16} \alpha^8 \hbar^5}{c^2} = \frac{e^{16} \alpha^8 \hbar^5}{\pi c^8 \varepsilon_0^3} = 64\pi^2 e^{16} \alpha^8 \hbar^5 \varepsilon_0 \mu_0 = \frac{e^{16} \alpha^8 \hbar^5 \mu_0}{\pi c^6 \varepsilon_0^2}$$

Planck angular frequency:

$$\omega_P = 1.855,103,785,054,537,578,662,092,235,839,7\text{e}43/\text{Sec.}$$

$$\omega_P = \frac{1}{t_P} = \sqrt{\frac{c^5}{\hbar G}} = \sqrt{8\pi}\, e^4\alpha^2\hbar = \frac{\sqrt{8\pi}\, e^8 c^2}{\hbar} = \frac{\sqrt{8\pi}\, e^8}{\hbar \varepsilon_0 \mu_0}$$

Planck pressure:

$$p_P' = 4.635,396,200,069,681,587,857,900,632,632\text{e}113\,\text{Pa}$$

$$p_P' = \frac{F_P}{l_P^2} = \frac{\hbar}{l_P^3 t_P} = \frac{c^7}{\hbar G^2} = \frac{64\pi^2 e^{16}\alpha^8 \hbar^5}{c^3} = \frac{e^{16}\alpha^8 \hbar^5}{\pi c^9 \varepsilon_0^3}$$

Planck force:

$$F_P = 1.210,575,396,641,860,935,039,016,403,887,3\text{e}44\,\text{Newton}$$

$$F_P = \frac{m_P l_P}{t_P^2} = \frac{\hbar c}{l_P^2} = \frac{q_P^2 c^2}{l_P^2} = \frac{8\pi e^8 \alpha^4 \hbar^3}{c} = \frac{\sqrt{\pi \varepsilon_0}\, e^8 \alpha^4 \hbar^3}{\pi c^4 \varepsilon_0^2} = \frac{e^8 \alpha^4 \hbar^3}{\sqrt{\pi \varepsilon_0}\, c^4 \varepsilon_0}$$

Planck current:

$$I_P' = 3.479,332,402,404,031,495,325,194,656,360,4\text{e}25\,\text{Amps}$$

$$I_P' = \sqrt{F_P} = \frac{q_P}{t_P} = \sqrt{\frac{\varepsilon_0 \mu_0 c^6}{G}} = \sqrt{8\pi\alpha}\, e^5 \alpha \hbar$$

Planck voltage:

$$V_P = 1.043,077,613,115,749,711,092,955,615,358,7\text{e}27\,\text{V}$$

$$V_P = \frac{E_P}{q_P} = \frac{\hbar}{t_P q_P} = \sqrt{\frac{c^4}{\varepsilon_0 \mu_0 G}} = \sqrt{8\pi\alpha}\, e^3 \alpha^2 \hbar^2$$

Planck impedance:

$$Z_P = \frac{V_P}{I_P'} = \frac{\hbar}{q_P^2} = \frac{1}{\varepsilon_0 \mu_0 c} = \frac{1}{\sqrt{\varepsilon_0 \mu_0}} = \frac{Z_0}{4\pi} = 29.979,245,8\,\Omega$$

Additional Structural Orientations:

$$\alpha\hbar = c\,e^2 = \frac{2h\alpha}{\mu_0} = \frac{e^2}{\sqrt{\varepsilon_0\mu_0}} = \frac{e^2}{c\,\varepsilon_0\,\mu_0} = 7.695{,}581{,}689{,}948{,}290{,}443{,}131{,}979{,}193{,}101{,}4$$

$$\frac{1}{\alpha\hbar} = \frac{1}{c\,e^2} = \frac{\mu_0}{2h\alpha} = \frac{\sqrt{\varepsilon_0\,\mu_0}}{e^2} = \frac{c\,\varepsilon_0\,\mu_0}{e^2} = 1.299{,}446{,}929{,}796{,}309{,}370{,}983{,}141{,}015{,}885{,}4$$

$$\alpha^2\hbar^2 = c^2 e^4 = \frac{e^2\alpha\hbar}{\sqrt{\varepsilon_0\mu_0}} = \frac{e^2\alpha\hbar}{c\varepsilon_0\mu_0} = \frac{e^4}{\varepsilon_0\mu_0} = 5.922{,}197{,}754{,}666{,}738{,}586{,}193{,}956{,}454{,}205{,}6$$

$$\frac{1}{\alpha^2\hbar^2} = \frac{1}{c^2 e^4} = \frac{\sqrt{\varepsilon_0\mu_0}}{e^2\alpha\hbar} = \frac{c\varepsilon_0\mu_0}{e^2\alpha\hbar} = \frac{\varepsilon_0\mu_0}{e^4} = 1.688{,}562{,}323{,}357{,}054{,}574{,}955{,}003{,}722{,}204{,}81$$

$$\alpha^4\hbar^4 = c^4 e^8 = \frac{e^4\alpha^2\hbar^2}{\varepsilon_0\mu_0} = \frac{e^8}{(\varepsilon_0\mu_0)^2} = 3.507{,}242{,}624{,}537{,}976{,}003{,}177{,}050{,}898{,}525{,}9$$

$$\frac{1}{\alpha^4\hbar^4} = \frac{1}{c^4 e^8} = \frac{\varepsilon_0\mu_0}{e^4\alpha^2\hbar^2} = \frac{(\varepsilon_0\mu_0)^2}{e^8} = 2.851{,}242{,}719{,}860{,}974{,}134{,}158{,}312{,}890{,}688{,}8$$

$$\alpha^8\hbar^8 = c^8 e^{16} = \frac{e^8\alpha^4\hbar^4}{(\varepsilon_0\mu_0)^2} = 1.230{,}075{,}082{,}737{,}603{,}011{,}435{,}284{,}283{,}985$$

$$\frac{1}{\alpha^8\hbar^8} = \frac{1}{c^8 e^{16}} = \frac{(\varepsilon_0\mu_0)^2}{e^8\alpha^4\hbar^4} = 8.129{,}585{,}047{,}560{,}205{,}424{,}273{,}714{,}105{,}489{,}7$$

$$e^2 c^2 = \alpha\hbar c = \frac{\alpha\hbar}{\sqrt{\varepsilon_0\mu_0}} = \frac{\alpha\hbar}{c\,\varepsilon_0\mu_0} = \frac{e^2}{\varepsilon_0\mu_0} = 2.307{,}077{,}350{,}569{,}391{,}884{,}844{,}445{,}260{,}704{,}7$$

$$\frac{1}{e^2 c^2} = \frac{1}{\alpha\hbar c} = \frac{\sqrt{\varepsilon_0\mu_0}}{\alpha\hbar} = \frac{c\,\varepsilon_0\mu_0}{\alpha\hbar} = \frac{\varepsilon_0\mu_0}{e^2} = 4.334{,}488{,}393{,}955{,}225{,}421{,}258{,}399{,}422{,}060{,}9$$

$$e^4 c^4 = \alpha^2\hbar^2 c^2 = \frac{\alpha^2\hbar^2}{\varepsilon_0\mu_0} = \frac{e^4}{(\varepsilon_0\mu_0)^2} = 5.322{,}605{,}901{,}510{,}284{,}741{,}921{,}062{,}987{,}100{,}2$$

$$\frac{1}{e^4 c^4} = \frac{1}{\alpha^2\hbar^2 c^2} = \frac{\varepsilon_0\mu_0}{\alpha^2\hbar^2} = \frac{(\varepsilon_0\mu_0)^2}{e^2} = 1.878{,}778{,}963{,}733{,}254{,}945{,}219{,}859{,}110{,}278$$

$$\frac{e^2}{\alpha} = \frac{\hbar}{c} = \frac{2h\varepsilon_0}{\sqrt{\varepsilon_0\mu_0}} = \frac{2h}{\mu_0 c} = \hbar\sqrt{\varepsilon_0\mu_0} = \hbar c\varepsilon_0\mu_0 = 3.517{,}672{,}633{,}845{,}832{,}886{,}163{,}209{,}257{,}441{,}7$$

$$\frac{\alpha}{e^2} = \frac{c}{\hbar} = \frac{\sqrt{\varepsilon_0\mu_0}}{2h\varepsilon_0} = \frac{\mu_0 c}{2h} = \frac{1}{\hbar\sqrt{\varepsilon_0\mu_0}} = \frac{1}{\hbar c\varepsilon_0\mu_0} = 2.842{,}788{,}696{,}078{,}040{,}026{,}200{,}517{,}986{,}812{,}7$$

$$\frac{e^4}{\alpha^2} = \frac{\hbar^2}{c^2} = \frac{\hbar\, e^2}{\alpha c} = \hbar^2\varepsilon_0\mu_0 = 1.237{,}402{,}075{,}890{,}787{,}908{,}121{,}088{,}442{,}422{,}2$$

$$\frac{\alpha^2}{e^4} = \frac{c^2}{\hbar^2} = \frac{\alpha c}{\hbar\, e^2} = \frac{1}{\hbar^2\varepsilon_0\mu_0} = 8.081{,}447{,}570{,}549{,}083{,}024{,}642{,}842{,}967{,}857{,}5$$

$$\frac{e^2}{\hbar} = \frac{\alpha}{c} = \frac{e^2\mu_0}{2h} = \alpha\sqrt{\varepsilon_0\mu_0} = \alpha c\varepsilon_0\mu_0 = 2.434{,}134{,}807{,}287{,}308{,}407{,}751{,}873{,}728{,}442{,}4$$

$$\frac{\hbar}{e^2} = \frac{c}{\alpha} = \frac{2h}{e^2\mu_0} = \frac{1}{\alpha\sqrt{\varepsilon_0\mu_0}} = \frac{1}{\alpha c\varepsilon_0\mu_0} = 4.108{,}235{,}899{,}697{,}098{,}939{,}265{,}503{,}086{,}827{,}3$$

$$\frac{e^4}{\hbar^2} = \frac{\alpha^2}{c^2} = \frac{\alpha\, e^2}{\hbar c} = \alpha^2\varepsilon_0\mu_0 = 5.925{,}012{,}260{,}047{,}622{,}040{,}387{,}715{,}044{,}905{,}8$$

$$\frac{\hbar^2}{e^4} = \frac{c^2}{\alpha^2} = \frac{\hbar c}{\alpha\, e^2} = \frac{1}{\alpha^2\varepsilon_0\mu_0} = 1.687{,}760{,}220{,}756{,}003{,}197{,}637{,}666{,}787{,}832{,}3$$

$$\frac{e^2}{c} = \alpha\hbar\varepsilon_0\mu_0 = e^2\sqrt{\varepsilon_0\mu_0} = e^2 c\varepsilon_0\mu_0 = 8.562{,}489{,}398{,}686{,}165{,}023{,}537{,}722{,}505{,}124{,}6$$

$$\frac{c}{e^2} = \frac{1}{\alpha\hbar\varepsilon_0\mu_0} = \frac{1}{e^2\sqrt{\varepsilon_0\mu_0}} = \frac{1}{e^2 c\varepsilon_0\mu_0} = 1.167{,}884{,}657{,}648{,}090{,}952{,}578{,}783{,}290{,}168{,}9$$

$$\frac{e^4}{c^2} = \alpha^2\hbar^2(\varepsilon_0\mu_0)^2 = e^4\varepsilon_0\mu_0 = 7.331{,}622{,}470{,}261{,}296{,}388{,}311{,}116{,}124{,}591{,}6$$

$$\frac{c^2}{e^4} = \frac{1}{\alpha^2\hbar^2(\varepsilon_0\mu_0)^2} = \frac{1}{e^4\varepsilon_0\mu_0} = 1.363{,}954{,}573{,}569{,}798{,}609{,}134{,}573{,}059{,}446{,}9$$

$$\frac{\alpha c}{e^2} = \frac{c^2}{\hbar} = \frac{\alpha}{e^2\sqrt{\varepsilon_0\mu_0}} = \frac{\alpha}{e^2 c\varepsilon_0\mu_0} = \frac{1}{2h\varepsilon_0} = \frac{1}{\hbar\varepsilon_0\mu_0} = 8.522,466,107,718,505,792,770,376,880,088,3$$

$$\frac{e^2}{\alpha c} = \frac{\hbar}{c^2} = \frac{e^2\sqrt{\varepsilon_0\mu_0}}{\alpha} = \frac{e^2 c\varepsilon_0\mu_0}{\alpha} = 2h\varepsilon_0 = \hbar\varepsilon_0\mu_0 = 1.173,369,289,312,085,658,326,737,911,946,3$$

$$\frac{\alpha^2 c^2}{e^4} = \frac{c^4}{\hbar^2} = \frac{e^4\varepsilon_0\mu_0}{\alpha^2} = \hbar^2(\varepsilon_0\mu_0)^2 = 1.376,795,489,100,748,975,133,279,397,667,8$$

$$\frac{e^4}{\alpha^2 c^2} = \frac{\hbar^2}{c^4} = \frac{\alpha^2}{e^4\varepsilon_0\mu_0} = \frac{1}{\hbar^2(\varepsilon_0\mu_0)^2} = 7.263,242,855,721,061,798,265,370,046,536,9$$

$$\frac{\hbar c}{e^2} = \frac{c^2}{\alpha} = \frac{\hbar}{e^2\sqrt{\varepsilon_0\mu_0}} = \frac{\hbar}{e^2 c\varepsilon_0\mu_0} = \frac{1}{\alpha\varepsilon_0\mu_0} = 1.231,618,138,414,034,746,471,597,884,985,6$$

$$\frac{e^2}{\hbar c} = \frac{\alpha}{c^2} = \frac{e^2\sqrt{\varepsilon_0\mu_0}}{\hbar} = \frac{e^2 c\varepsilon_0\mu_0}{\hbar} = \alpha\varepsilon_0\mu_0 = 8.119,399,745,831,192,350,248,763,524,544$$

$$\frac{\hbar^2 c^2}{e^4} = \frac{c^4}{\alpha^2} = \frac{\hbar^2}{e^4\varepsilon_0\mu_0} = \frac{1}{\alpha^2(\varepsilon_0\mu_0)^2} = 1.516,883,238,870,452,451,204,727,685,318,7$$

$$\frac{e^4}{\hbar^2 c^2} = \frac{\alpha^2}{c^4} = \frac{e^4\varepsilon_0\mu_0}{\hbar^2} = \alpha^2(\varepsilon_0\mu_0)^2 = 6.592,465,223,260,363,093,900,240,321,636,6$$

$$\alpha c = \frac{c^2 e^2}{\hbar} = \frac{e^2}{\hbar\varepsilon_0\mu_0} = \frac{\alpha}{\sqrt{\varepsilon_0\mu_0}} = \frac{\alpha}{c\varepsilon_0\mu_0} = 2.187,691.263,793,014,029,297,221,235,591,6$$

$$\frac{1}{\alpha c} = \frac{\hbar}{c^2 e^2} = \frac{\hbar\varepsilon_0\mu_0}{e^2} = \frac{\sqrt{\varepsilon_0\mu_0}}{\alpha} = \frac{c\varepsilon_0\mu_0}{\alpha} = 4.571,028,904,079,464,685,754,372,971,864,4$$

$$\alpha^2 c^2 = \frac{c^4 e^4}{\hbar^2} = \frac{e^4}{\hbar^2(\varepsilon_0\mu_0)^2} = \frac{\varepsilon_0\mu_0}{\alpha^2} = 4.785,993,065,676,274,896,288,785,105,225,9$$

$$\frac{1}{\alpha^2 c^2} = \frac{\hbar^2}{c^4 e^4} = \frac{\hbar^2(\varepsilon_0\mu_0)^2}{e^4} = \frac{\alpha^2}{\varepsilon_0\mu_0} = 2.089,430,524,192,991,196,686,734,643,385,9$$

$$\hbar c = \frac{c^2 e^2}{\alpha} = \frac{e^2}{\alpha \varepsilon_0 \mu_0} = \frac{\hbar}{\sqrt{\varepsilon_0 \mu_0}} = \frac{\hbar}{c \varepsilon_0 \mu_0} = 3.161{,}526{,}496{,}769{,}723{,}608{,}524{,}739{,}584{,}240{,}4$$

$$\frac{1}{\hbar c} = \frac{\alpha}{c^2 e^2} = \frac{\alpha \varepsilon_0 \mu_0}{e^2} = \frac{\sqrt{\varepsilon_0 \mu_0}}{\hbar} = \frac{c \varepsilon_0 \mu_0}{\hbar} = 3.163{,}029{,}002{,}039{,}824{,}088{,}408{,}177{,}844{,}831{,}4$$

$$\hbar^2 c^2 = \frac{c^4 e^4}{\alpha^2} = \frac{\alpha^2 (\varepsilon_0 \mu_0)^2}{e^4} = \frac{\hbar^2}{\varepsilon_0 \mu_0} = 9.995{,}249{,}789{,}777{,}041{,}182{,}487{,}865{,}767{,}672{,}7$$

$$\frac{1}{\hbar^2 c^2} = \frac{\alpha^2}{c^4 e^4} = \frac{e^4}{\alpha^2 (\varepsilon_0 \mu_0)^2} = \frac{\varepsilon_0 \mu_0}{\hbar^2} = 1.000{,}475{,}246{,}774{,}504{,}549{,}722{,}814{,}303{,}236{,}8$$

$$\frac{\alpha}{\hbar e^2} = \frac{\alpha^2 \sqrt{\varepsilon_0 \mu_0}}{e^4} = \frac{\alpha^2 c \varepsilon_0 \mu_0}{e^4} = \frac{1}{\hbar^2 \sqrt{\varepsilon_0 \mu_0}} = \frac{1}{\hbar^2 c \varepsilon_0 \mu_0} = 2.695{,}680{,}746{,}761{,}508{,}931{,}836{,}718{,}510{,}11$$

$$\frac{\hbar e^2}{\alpha} = \frac{e^4}{\alpha^2 \sqrt{\varepsilon_0 \mu_0}} = \frac{e^4}{\alpha^2 c \varepsilon_0 \mu_0} = \hbar^2 \sqrt{\varepsilon_0 \mu_0} = \hbar^2 c \varepsilon_0 \mu_0 = 3.709{,}638{,}098{,}656{,}018{,}465{,}322{,}992{,}657{,}832$$

$$\frac{\alpha^2}{\hbar^2 e^4} = \frac{\alpha^4 \varepsilon_0 \mu_0}{e^8} = \frac{1}{\hbar^4 \varepsilon_0 \mu_0} = 7.266{,}694{,}688{,}460{,}686{,}447{,}498{,}432{,}860{,}292{,}9$$

$$\frac{\hbar^2 e^4}{\alpha^2} = \frac{e^8}{\hbar^4 \varepsilon_0 \mu_0} = \hbar^4 \varepsilon_0 \mu_0 = 1.376{,}141{,}482{,}300{,}023{,}978{,}833{,}769{,}110{,}458{,}3$$

$$\frac{\hbar}{\alpha e^2} = \frac{\hbar^2 \sqrt{\varepsilon_0 \mu_0}}{e^4} = \frac{\hbar^2 c \varepsilon_0 \mu_0}{e^4} = \frac{1}{\alpha^2 \sqrt{\varepsilon_0 \mu_0}} = \frac{1}{\alpha^2 c \varepsilon_0 \mu_0} = 5.629{,}762{,}109{,}479.095{,}693{,}720{,}443{,}053{,}415$$

$$\frac{\alpha e^2}{\hbar} = \frac{e^4}{\hbar^2 \sqrt{\varepsilon_0 \mu_0}} = \frac{e^4}{\hbar^2 c \varepsilon_0 \mu_0} = \alpha^2 \sqrt{\varepsilon_0 \mu_0} = \alpha^2 c \varepsilon_0 \mu_0 = 1.776{,}273{,}989{,}119{,}811{,}808{,}542{,}808{,}366{,}315$$

$$\frac{\hbar^2}{\alpha^2 e^4} = \frac{\hbar^4 \varepsilon_0 \mu_0}{e^8} = \frac{1}{\alpha^4 \varepsilon_0 \mu_0} = 3.169{,}422{,}140{,}932{,}651{,}744{,}741{,}437{,}176{,}535{,}8$$

$$\frac{\alpha^2 e^4}{\hbar^2} = \frac{e^8}{\hbar^4 \varepsilon_0 \mu_0} = \alpha^4 \varepsilon_0 \mu_0 = 3.155{,}149{,}284{,}423{,}609{,}319{,}193{,}631{,}857{,}663{,}3$$

$$\alpha^2 c = \frac{\alpha^3 \hbar}{e^2} = \frac{\alpha^2}{\sqrt{\varepsilon_0 \mu_0}} = \frac{\alpha^2}{c\,\varepsilon_0 \mu_0} = 1.596{,}435{,}446{,}576{,}936{,}533{,}970{,}039{,}067{,}902{,}8$$

$$\frac{1}{\alpha^2 c} = \frac{e^2}{\alpha^3 \hbar} = \frac{\sqrt{\varepsilon_0 \mu_0}}{\alpha^2} = \frac{c\,\varepsilon_0 \mu_0}{\alpha^2} = 6.263{,}955{,}126{,}680{,}452{,}972{,}270{,}776{,}347{,}344{,}3$$

$$\alpha^4 c^2 = \frac{\alpha^6 \hbar^2}{e^4} = \frac{\alpha^4}{\varepsilon_0 \mu_0} = 2.548{,}606{,}135{,}087{,}302{,}782{,}177{,}162{,}704{,}238{,}9$$

$$\frac{1}{\alpha^4 c^2} = \frac{e^4}{\alpha^6 \hbar^2} = \frac{\varepsilon_0 \mu_0}{\alpha^4} = 3.923{,}713{,}382{,}906{,}632{,}964{,}377{,}794{,}696{,}936{,}6$$

$$\hbar^2 c = \frac{\hbar^3 \alpha}{e^2} = \frac{\hbar^2}{\sqrt{\varepsilon_0 \mu_0}} = \frac{\hbar^2}{c\,\varepsilon_0 \mu_0} = 3.334{,}056{,}452{,}406{,}498{,}225{,}678{,}467{,}791{,}098{,}5$$

$$\frac{1}{\hbar^2 c} = \frac{e^2}{\hbar^3 \alpha} = \frac{\sqrt{\varepsilon_0 \mu_0}}{\hbar^2} = \frac{c\,\varepsilon_0 \mu_0}{\hbar^2} = 2.999{,}349{,}333{,}986{,}852{,}906{,}935{,}857{,}186{,}449{,}3$$

$$\hbar^4 c^2 = \frac{\hbar^6 \alpha^2}{e^4} = \frac{\hbar^4}{\varepsilon_0 \mu_0} = = 1.111{,}593{,}242{,}783{,}340{,}436{,}826{,}493{,}604{,}025{,}2$$

$$\frac{1}{\hbar^4 c^2} = \frac{e^4}{\hbar^6 \alpha^2} = \frac{\varepsilon_0 \mu_0}{\hbar^4} = = 8.996{,}096{,}427{,}287{,}378{,}106{,}348{,}227{,}311{,}847{,}5$$

$$\alpha^2 \hbar = \alpha c e^2 = \frac{\alpha e^2}{\sqrt{\varepsilon_0 \mu_0}} = \frac{\alpha e^2}{c\,\varepsilon_0 \mu_0} = 5.615{,}737{,}282{,}125{,}140{,}775{,}875{,}703{,}364{,}458{,}5$$

$$\frac{1}{\alpha^2 \hbar} = \frac{1}{\alpha c e^2} = \frac{\sqrt{\varepsilon_0 \mu_0}}{\alpha e^2} = \frac{c\,\varepsilon_0 \mu_0}{\alpha e^2} = 1.780{,}710{,}082{,}686{,}727{,}893{,}593{,}597{,}596{,}406{,}9$$

$$\alpha^4 \hbar^2 = \alpha^2 c^2 e^4 = \frac{\alpha^2 e^4}{\varepsilon_0 \mu_0} = 3.153{,}650{,}522{,}185{,}026{,}296{,}518{,}284{,}737{,}731{,}2$$

$$\frac{1}{\alpha^4 \hbar^2} = \frac{1}{\alpha^2 c^2 e^4} = \frac{\varepsilon_0 \mu_0}{\alpha^2 e^4} = 3.170{,}928{,}398{,}582{,}173{,}291{,}897{,}080{,}100{,}985$$

$$\alpha \hbar^2 = \hbar c e^2 = \frac{\hbar e^2}{\sqrt{\varepsilon_0 \mu_0}} = \frac{\hbar e^2}{c \varepsilon_0 \mu_0} = 8.115,542,860,263,498,694,803,471,418,117,4$$

$$\frac{1}{\alpha \hbar^2} = \frac{1}{\hbar c e^2} = \frac{\sqrt{\varepsilon_0 \mu_0}}{\hbar e^2} = \frac{c \varepsilon_0 \mu_0}{\hbar e^2} = 1.232,203,460,961,737,314,551,532,289,172,9$$

$$\alpha^2 \hbar^4 = \hbar^2 c^2 e^4 = \frac{\hbar^2 e^4}{\varepsilon_0 \mu_0} = 6.586,203,591,677,384,950,253,269,470,319,7$$

$$\frac{1}{\alpha^2 \hbar^4} = \frac{1}{\hbar^2 c^2 e^4} = \frac{\varepsilon_0 \mu_0}{\hbar^2 e^4} = 1.518,325,369,206,083,694,127,951,532,241,7$$

RELATIONSHIPS OF CONSTANTS :

LEGEND:

$\alpha = \frac{c e^2}{\hbar}$ Fine structure constant = 7.297,352,569,800,184,997,640,004,791,553,6e-3

$c = \frac{\alpha \hbar}{e^2}$ Speed of light = 299,792,458 meters per second

$e^2 = \frac{\alpha \hbar}{c}$ Elementary Charge = 2.566,969,743,431,067,383,000,001,685,533,2e-38C

$\hbar = \frac{c e^2}{\alpha}$ Reduced Planck Constant 1.054,571,725,339,976,234,000,102,692,456,8e-34 J·S

ε_0 Permittivity Constant = 8.854,187,817,620,389,850,536,563,031,710,9e-12 Farads/meter

$A_r(e)$ Relative mass of an electron 5.485,799,109,910,795,715,106,472,125,503,7e-4 Kg

N_A Avogadro Number   = 6.022,141,310,907,101,425,969,05 Mol^{-1}

K_B Boltzmann Constant   = 1.380,648,798,394,132,924,292,774,930,237,3e-23 J/K

m_e Mass of an electron   = 9.109,382,903,345,524,937,498,006,139,342,6e-31 Kg

m_u atomic mass unit   = 1.660,538,915,267,287,672,145,963,619,188,4e-27 Kg

λ_e electron wavelength   = 2.426,310,238,899,999,814,307,976,139,616,8e-12 meter

r_e classical electron radius = 2.817,940,326,658,481,354,953,560,028,397,4e-15 meter

R   Gas Constant   = 8.314,462,164,663,558,038,800,395,847,263,8 J/Mol·K

R_∞ Rydberg Constant   = 1.097,373,156,866,196,496,466,432,160,158,9E7/m

FORMULAS:

4.798,310,356,941,517,601,264,353,822,911,1 =

$$= \frac{\sqrt{\pi\varepsilon_0}\,\hbar\,A_r(e)\lambda_e}{\pi\alpha^2 c^3 \varepsilon_0^2 m_u N_A K_B} = \frac{\sqrt{\pi\varepsilon_0}\,\hbar\,A_r(e)\lambda_e}{\pi\alpha^2 c^3 \varepsilon_0^2 m_u R} = \frac{e^2 \hbar N_A}{\alpha^3 c^4 \varepsilon_0^2 m_u R} = \frac{\hbar\,A_r(e)\lambda_e}{\sqrt{\pi\varepsilon_0}\,\alpha^2 c^3 \varepsilon_0 m_u N_A K_B}$$

$$= \frac{\hbar\,A_r(e)\lambda_e}{\sqrt{\pi\varepsilon_0}\,\alpha^2 c^3 \varepsilon_0 m_u R} = \frac{\hbar^2}{\alpha^2 c^5 \varepsilon_0^2 m_u K_B} = \frac{\hbar^2 N_A}{\alpha^2 c^5 \varepsilon_0^2 m_u R} = \frac{e^2 \hbar}{\alpha^3 c^4 \varepsilon_0^2 m_u K_B} =$$

$$\frac{\sqrt{\pi\varepsilon_0}\,\hbar\,A_r(e)^2 \lambda_e}{\pi\alpha^2 c^3 \varepsilon_0^2 m_e N_A K_B} = \frac{\sqrt{\pi\varepsilon_0}\,\hbar\,A_r(e)^2 \lambda_e}{\pi\alpha^2 c^3 \varepsilon_0^2 m_e R} = \frac{\hbar^2 A_r(e)}{\alpha^2 c^5 \varepsilon_0^2 m_e K_B} = \frac{\hbar^2 A_r(e) N_A}{\alpha^2 c^5 \varepsilon_0^2 m_e R} = \frac{e^2 \hbar\,A_r(e)}{\alpha^3 c^4 \varepsilon_0^2 m_e K_B}$$

$$= \frac{e^2 \hbar\,A_r(e) N_A}{\alpha^3 c^4 \varepsilon_0^2 m_e R} = \frac{\sqrt{\pi\varepsilon_0}\,A_r(e)^2 \lambda_e}{\pi\varepsilon_0 R_\infty N_A K_B} = \frac{A_r(e)^2 \lambda_e}{\pi\varepsilon_0 R_\infty R} = \frac{\sqrt{\pi\varepsilon_0}\,\hbar\,A_r(e)}{c^2 \varepsilon_0 R_\infty K_B} = \frac{\hbar\,A_r(e) N_A}{c^2 \varepsilon_0 R_\infty R} =$$

$$\frac{e^2 A_r(e)}{\alpha c \varepsilon_0 R_\infty K_B} = \frac{e^2 A_r(e) N_A}{\alpha c \varepsilon_0 R_\infty R} = \frac{\sqrt{\pi\varepsilon_0}\,\hbar r_e A_r(e)^2 \lambda_e}{\pi e^2 \alpha^2 c^3 \varepsilon_0^2 N_A K_B} = \frac{\sqrt{\pi\varepsilon_0}\,\hbar r_e A_r(e)^2 \lambda_e}{\pi e^2 \alpha^2 c^3 \varepsilon_0^2 R} = \frac{\hbar^2 r_e A_r(e)}{e^2 \alpha^2 c^5 \varepsilon_0^2 K_B}$$

$$= \frac{\hbar^2 r_e A_r(e) N_A}{e^2 \alpha^2 c^5 \varepsilon_0^2 R} = \frac{\hbar r_e A_r(e)}{\alpha^3 c^4 \varepsilon_0^2 K_B} = \frac{\hbar r_e A_r(e) N_A}{\alpha^3 c^4 \varepsilon_0^2 R} = \frac{\hbar r_e A_r(e)}{\alpha^3 c^4 \varepsilon_0^2 K_B} = \frac{\sqrt{\pi\varepsilon_0}\,r_e A_r(e)^2 \lambda_e}{\pi\alpha^3 c^2 \varepsilon_0^2 N_A K_B} =$$

$$\frac{\sqrt{\pi\varepsilon_0}\,r_e A_r(e)^2 \lambda_e}{\pi\alpha^3 c^2 \varepsilon_0^2 R} = \frac{e^2 r_e A_r(e) N_A}{\alpha^4 c^3 \varepsilon_0^2 R} = \frac{e^2 r_e A_r(e)}{\alpha^4 c^3 \varepsilon_0^2 K_B} = \frac{\hbar\,A_r(e)^2 \lambda_e}{\sqrt{\pi\varepsilon_0}\,\alpha^2 c^3 \varepsilon_0 m_e N_A K_B} =$$

$$\frac{\hbar\,A_r(e)^2 \lambda_e}{\sqrt{\pi\varepsilon_0}\,\alpha^2 c^3 \varepsilon_0 m_e R} = \frac{\hbar^2 A_r(e)}{\alpha^2 c^5 \varepsilon_0^2 m_e K_B} = \frac{\hbar^2 A_r(e) N_A}{\alpha^2 c^5 \varepsilon_0^2 m_e R} = \frac{e^2 \hbar\,A_r(e)}{\alpha^3 c^4 \varepsilon_0^2 m_e K_B} = \frac{e^2 \hbar\,A_r(e) N_A}{\alpha^3 c^4 \varepsilon_0^2 m_e R} =$$

$$\frac{A_r(e)^2 \lambda_e}{\sqrt{\pi\varepsilon_0}\,R_\infty N_A K_B} = \frac{A_r(e)^2 \lambda_e}{\sqrt{\pi\varepsilon_0}\,R_\infty R} = \frac{\hbar A_r(e)}{c^2 \varepsilon_0 R_\infty K_B} = \frac{\hbar A_r(e) N_A}{c^2 \varepsilon_0 R_\infty R} = \frac{e^2 A_r(e)}{\alpha c \varepsilon_0 R_\infty K_B} = \frac{e^2 A_r(e) N_A}{\alpha c \varepsilon_0 R_\infty R}$$

$$= \frac{\hbar\,r_e A_r(e)^2 \lambda_e}{\sqrt{\pi\varepsilon_0}\,e^2 \alpha^2 c^3 \varepsilon_0 N_A K_B} = \frac{\hbar\,r_e A_r(e) N_A}{\alpha^3 c^4 \varepsilon_0^2 R} = \frac{\hbar\,r_e A_r(e)^2 \lambda_e}{\sqrt{\pi\varepsilon_0}\,e^2 \alpha^2 c^3 \varepsilon_0 R} = \frac{\hbar^2 r_e A_r(e)}{e^2 \alpha^2 c^5 \varepsilon_0^2 K_B} =$$

$$\frac{\hbar^2 r_e A_r(e) N_A}{e^2 \alpha^2 c^5 \varepsilon_0^2 R} = \frac{\hbar r_e A_r(e)}{\alpha^3 c^4 \varepsilon_0^2 K_B} = \frac{r_e A_r(e)^2 \lambda_e}{\sqrt{\pi\varepsilon_0}\,\alpha^3 c^2 \varepsilon_0 N_A K_B} = \frac{r_e A_r(e)^2 \lambda_e}{\sqrt{\pi\varepsilon_0}\,\alpha^3 c^2 \varepsilon_0 R} = \frac{\hbar r_e A_r(e)}{\alpha^3 c^4 \varepsilon_0^2 K_B} =$$

$$\frac{\hbar r_e A_r(e) N_A}{\alpha^3 c^4 \varepsilon_0^2 R} = \frac{e^2 r_e A_r(e)}{\alpha^4 c^3 \varepsilon_0^2 K_B} = \frac{e^2 r_e A_r(e) N_A}{\alpha^4 c^3 \varepsilon_0^2 R}$$

QUANTUM JUMP NUMBER

Under certain circumstances, quantum constants and/or the math they produce can differ by a certain arbitrary value which is listed below. This value first appeared in the mathematics regarding these physical constants and appears quite frequently in those particular mathematics and relationships. At first it was suspected that perhaps it was a quirk in the computer or the calculator which was being used, but different computers and calculators were tried with the same result.

For instance, the product of $\alpha \hbar$ divided by the product of $c e^2$ can yield this particular value which is thought to represent the initial quantum jump in the primordial singularity that produced the existence of our universe and the birth of four-dimensional space-time. This number has been named a "quantum jump number" and it appears quite frequently in the mathematics associated with the physical constants related to atoms and electromagnetic waves.

When you divide the numerical values which are represented by the numerator and denominator in these formulas directly, you get "1" or unity. However, as these formulas are derived down from higher numbers and formulas, you may get the following numerical value for each of these particular formulas. This process may quite possibly provide us with a detailed analysis of how the physical expansion process of the primordial singularity actually occurred, and how all other quantum changes-of-state and dual existences are able to take place.

Here are some examples:

$$\frac{e^2 c}{\alpha \hbar} = \frac{\alpha \hbar}{e^2 c} = \frac{e^4 c^2}{\alpha^2 \hbar^2} = \frac{\alpha^2 \hbar^2}{e^4 c^2} = .000,000,000,000,006,537,188,612,504,934,7$$

$$= \frac{\alpha^2 \hbar^6}{e^{14} c^7} = \frac{e^{14} c^7}{\alpha^2 \hbar^6} = \frac{\alpha^3 \hbar^3}{e^6 c^3} = \frac{e^6 c^3}{\alpha^3 \hbar^3} = \frac{\alpha^3 \hbar^5}{e^6 c} = \frac{e^6 c}{\alpha^3 \hbar^5} = \frac{\alpha^4 \hbar^4}{e^8 c^4} = \frac{e^8 c^4}{\alpha^4 \hbar^4}$$

$$= \frac{\alpha^5 \hbar^5}{e^{10} c^5} = \frac{e^{10} c^5}{\alpha^5 \hbar^5} = \frac{\alpha^5 \hbar^6}{e^{10} c^8} = \frac{e^{10} c^8}{\alpha^5 \hbar^6} = \frac{\alpha^5 \hbar^7}{e^{10} c^3} = \frac{e^{10} c^3}{\alpha^5 \hbar^7} = \frac{e^{12} c^6}{\alpha^6 \hbar^6} = \frac{\alpha^6 \hbar^6}{e^{12} c^6}$$

$$= \frac{\sqrt{2c}}{\sqrt{\pi\varepsilon_0}\sqrt{8c}\, c} \wedge = \frac{\sqrt{\pi\varepsilon_0}\sqrt{8c}\, c}{\sqrt{2c}} \wedge = \wedge \frac{\sqrt{8\pi}\sqrt{8c}\, c\varepsilon_0}{\sqrt{8\varepsilon_0}\sqrt{2c}} = \frac{\sqrt{8\varepsilon_0}\sqrt{8c}\, \pi c}{\sqrt{8\pi}\sqrt{2c}} \wedge = \frac{\pi c \varepsilon_0}{(\sqrt{\pi\varepsilon_0} - \pi c \varepsilon_0)}$$

$$= \frac{\alpha^4 \hbar^4}{(\pi c \varepsilon_0 - \sqrt{\pi\varepsilon_0})\sqrt{\pi\varepsilon_0}\, e^8 c^6} = \frac{(\pi c^3 \varepsilon_0 - c)}{(\sqrt{\pi\varepsilon_0} + \pi c \varepsilon_0)\sqrt{\pi\varepsilon_0}\, c^3} = \frac{\sqrt{8c}\sqrt{2c}\, \pi \alpha^4 \hbar^4 \varepsilon_0}{(\pi c \varepsilon_0 - \sqrt{\pi\varepsilon_0})\sqrt{\pi\varepsilon_0}\, e^8 c^5}$$

$$= \frac{\sqrt{8\pi}\, \alpha^4 \hbar^4}{(\pi c \varepsilon_0 - \sqrt{\pi\varepsilon_0})\sqrt{8\varepsilon_0}\, \pi e^8 c^6} = \frac{\sqrt{8\varepsilon_0}\, \alpha^4 \hbar^4}{(\pi c \varepsilon_0 - \sqrt{\pi\varepsilon_0})\sqrt{8\pi}\, e^8 c^6 \varepsilon_0} = \frac{\sqrt{8\pi}\, e^6}{\sqrt{8\varepsilon_0}\sqrt{8c}\sqrt{2c}\, \pi \alpha \hbar^2 c \varepsilon_0 \lambda_p m_u}$$

$$= \frac{\sqrt{8\varepsilon_0}}{(\pi c \varepsilon_0 - \sqrt{\pi\varepsilon_0})\sqrt{8\pi}\, c^2 \varepsilon_0} = \frac{\sqrt{8\pi}}{(\pi c \varepsilon_0 - \sqrt{\pi\varepsilon_0})\sqrt{8\varepsilon_0}\, \pi c^2} = \frac{\sqrt{8\pi}\sqrt{8c}\, \alpha^4 \hbar^3 \varepsilon_0^2 \lambda_p m_u}{\sqrt{\pi\varepsilon_0}\sqrt{8\varepsilon_0}\sqrt{2c}\, e^8 c}$$

$$= \frac{\sqrt{8c}\, \pi e^{20} c^{11} \varepsilon_0}{\sqrt{\pi\varepsilon_0}\sqrt{2c}\, \alpha^{10} \hbar^{10}} = \frac{\sqrt{8c}\, \pi \alpha^4 \hbar^4 \varepsilon_0}{\sqrt{\pi\varepsilon_0}\sqrt{2c}\, e^8 c^3} = \frac{\sqrt{\pi\varepsilon_0}\sqrt{8c}\, e^{20} c^{11}}{\sqrt{2c}\, \alpha^{10} \hbar^{10}} = \frac{\sqrt{\pi\varepsilon_0}\sqrt{8c}\, \alpha^4 \hbar^4}{\sqrt{2c}\, e^8 c^3}$$

$$= \frac{\pi \hbar}{\sqrt{\pi\varepsilon_0}\, c^2 \lambda_p m_u} \wedge = \frac{\pi e^2}{\sqrt{\pi\varepsilon_0}\, \alpha c \lambda_p m_u} = \frac{\sqrt{8\varepsilon_0}\, \pi}{\sqrt{\pi\varepsilon_0}\sqrt{8\pi}} = \frac{\sqrt{\pi\varepsilon_0}\sqrt{8\pi}}{\sqrt{8\varepsilon_0}\, \pi} = \frac{\sqrt{8\pi}\, \varepsilon_0}{\sqrt{\pi\varepsilon_0}\sqrt{8\varepsilon_0}} = \frac{\sqrt{\pi\varepsilon_0}\sqrt{8\varepsilon_0}}{\sqrt{8\pi}\, \varepsilon_0}$$

$$= \wedge \frac{\sqrt{8\varepsilon_0}\, c^2 \lambda_p m_e}{\sqrt{8\pi}\, \hbar\, A_r(e)} = \wedge \frac{\sqrt{8\pi}\, e^2 c^3 \varepsilon_0 \lambda_p m_u}{\sqrt{8\varepsilon_0}\, \pi \alpha \hbar^2} = \wedge \frac{\sqrt{8\pi}\, c^2 \varepsilon_0 \lambda_p m_u}{\sqrt{8\varepsilon_0}\, \pi \hbar} = \frac{\sqrt{8\pi}\, e^6}{\sqrt{8\varepsilon_0}\sqrt{8c}\sqrt{2c}\, \pi \alpha \hbar^2 c \varepsilon_0 \lambda_e m_e}$$

In the equations below, it is the **difference** between the numbers in parentheses which yields the quantum jump number. The ^ or v indicators determine which number in parentheses that particular equation represents. The equations which lack the up or down indicators have not yet been checked for accuracy. That might challenge some of the readers to assist in re-checking this math. Most of the math in this book is accurate to 32 decimal places.

$$\frac{\varepsilon_0}{\sqrt{\pi\varepsilon_0}\,\pi e^6\hbar^4 c^4} \wedge = \frac{\varepsilon_0}{\sqrt{\pi\varepsilon_0}\,\pi\alpha^3\hbar^7 c} \wedge = *\begin{cases}1.000,000,006,661,624,754,502,248,052,661,4\\ 1.000,000,006,661,618,217,313,591,999,471,7\end{cases}$$

$$= \wedge \frac{\varepsilon_0}{\sqrt{\pi\varepsilon_0}\,\pi e^4\alpha\hbar^5 c^3} = \wedge \frac{\varepsilon_0}{\sqrt{\pi\varepsilon_0}\,\pi e^2\alpha^2\hbar^6 c^2} = \wedge \frac{1}{\pi\alpha^3\hbar^6 c^3\, m_e\lambda_e} = \wedge \frac{\sqrt{8c}\,\sqrt{2c}\,\varepsilon_0}{(\pi c\varepsilon_0 - \sqrt{\pi\varepsilon_0})\,\pi e^6\hbar^4 c^5}$$

$$= \wedge \frac{1}{\pi\alpha^3\hbar^6 c^3\, \lambda_p m_u} = \wedge \frac{1}{(\pi c\varepsilon_0 - \sqrt{\pi\varepsilon_0})\pi^2 e^6\hbar^4 c^6} = \wedge \frac{\sqrt{\pi\varepsilon_0}}{\pi^2 e^2\alpha^2\hbar^6 c^2} = \wedge \frac{\sqrt{\pi\varepsilon_0}}{\pi^2 e^4\alpha\hbar^5 c^3}$$

$$= \wedge \frac{\sqrt{\pi\varepsilon_0}}{\pi^2 e^6\hbar^4 c^4} = \wedge \frac{\sqrt{\pi\varepsilon_0}}{\pi^2\alpha^3\hbar^7 c} = v \frac{\sqrt{\pi\varepsilon_0}\, e^2}{\pi^2\alpha^4\hbar^8} = \wedge \frac{\sqrt{\pi\varepsilon_0}\,\alpha}{\pi^2 e^8\hbar^3 c^5} = \wedge \frac{\sqrt{\pi\varepsilon_0}\,\alpha^2}{\pi^2 e^{10}\hbar^2 c^6} = \wedge \frac{\alpha^5\hbar\varepsilon_0}{\sqrt{\pi\varepsilon_0}\,\pi e^{16} c^9}$$

$$= \wedge \frac{\alpha\varepsilon_0}{\sqrt{\pi\varepsilon_0}\,\pi e^8\hbar^3 c^5} = \wedge \frac{e^6 c^2\varepsilon_0}{\sqrt{\pi\varepsilon_0}\,\pi\alpha^6\hbar^{10}} = \frac{\alpha^2\varepsilon_0}{\sqrt{\pi\varepsilon_0}\,\pi e^{10}\hbar^2 c^6} = v \frac{e^2\varepsilon_0}{\sqrt{\pi\varepsilon_0}\,\pi\alpha^4\hbar^8} = \frac{\sqrt{\pi\varepsilon_0}\, e^4 c}{\pi^2\alpha^7\hbar^7}$$

$$= \frac{c^2\varepsilon_0}{\sqrt{\pi\varepsilon_0}\,\pi e^2\alpha^6\hbar^{10}} = \frac{\alpha\varepsilon_0}{\sqrt{\pi\varepsilon_0}\,\pi e^{10}\hbar^2 c^6} = \frac{\alpha^4\varepsilon_0}{\sqrt{\pi\varepsilon_0}\,\pi e^{14} c^8} = \frac{\alpha^3\varepsilon_0}{\sqrt{\pi\varepsilon_0}\,\pi e^{14} c^8} = \frac{\alpha^2\varepsilon_0^3}{\sqrt{\pi\varepsilon_0}\,\pi e^{16}\hbar^2 c^4}$$

$$= \frac{\sqrt{\pi\varepsilon_0}\,\pi^3\hbar^8 c^3\varepsilon_0}{e^2\alpha^2 c^5} = \frac{\sqrt{\pi\varepsilon_0}\,\pi^3 e^8\hbar^3 c^3\varepsilon_0}{\alpha^7} = \frac{e^{14}\hbar^4 c^8\varepsilon_0}{\sqrt{\pi\varepsilon_0}\,\pi} = \frac{\sqrt{\pi\varepsilon_0}\, e^{14}\hbar^4 c^8}{\pi^2} = \frac{e^4 c\varepsilon_0}{\sqrt{\pi\varepsilon_0}\,\pi\alpha^7\hbar^7}$$

$$= \frac{\sqrt{\pi\varepsilon_0}\, e^8 c^3}{\alpha^7\hbar^7} = \frac{\sqrt{\pi\varepsilon_0}\, e^6 c^2}{\pi^2\alpha^6\hbar^{10}} = \frac{\sqrt{\pi\varepsilon_0}\, e^6 c^2}{\alpha^6\hbar^6} = \wedge \frac{\sqrt{\pi\varepsilon_0}\, e^4 c}{\pi^2\alpha^5\hbar^9} = \frac{\pi e^8 c^3\varepsilon_0}{\sqrt{\pi\varepsilon_0}\,\alpha^7\hbar^7} = \frac{\pi e^4 c}{\sqrt{\pi\varepsilon_0}\,\alpha^5\hbar^5}$$

$$= \frac{\pi e^8}{\sqrt{\pi\varepsilon_0}\,\alpha^3\hbar^7 c} = \frac{\sqrt{\pi\varepsilon_0}\, e^{12} c}{\alpha^5\hbar^9} = \frac{\pi e^{12} c\varepsilon_0}{\sqrt{\pi\varepsilon_0}\,\alpha^5\hbar^9} = \frac{\pi e^6\varepsilon_0}{\sqrt{\pi\varepsilon_0}\,\alpha^2\hbar^6 c^2} = \frac{\sqrt{\pi\varepsilon_0}\, e^8\alpha^5}{\alpha^8\hbar^7 c}$$

$$= \frac{(\pi c\varepsilon_0 - \sqrt{\pi\varepsilon_0})\varepsilon_0}{\pi e^2\alpha^2\hbar^6} = \frac{c\varepsilon_0^2\lambda_e^2 m_e^2}{\sqrt{\pi\varepsilon_0}\,\pi^3 e^4\alpha\hbar^6} = \frac{\alpha\varepsilon_0^2\lambda_e^2 m_e^2}{\sqrt{\pi\varepsilon_0}\,\pi^3 e^8\hbar^4 c} = \frac{e^2 c^4\varepsilon_0^2\lambda_e^2 m_e^2}{\sqrt{\pi\varepsilon_0}\,\pi^2\alpha^4\hbar^{10}}$$

$$= \frac{\sqrt{8\varepsilon_0}\,\alpha\varepsilon_0\,\lambda_p m_u}{\sqrt{\pi\varepsilon_0}\,\sqrt{8\pi}\,\pi e^4\alpha^2\hbar^6 c} = \frac{\sqrt{8\varepsilon_0}\,\alpha\varepsilon_0\,\lambda_p m_u}{\sqrt{\pi\varepsilon_0}\,\sqrt{8\pi}\,\pi e^2\alpha^3\hbar^7} = \frac{\sqrt{8\varepsilon_0}\,\alpha\varepsilon_0\,\lambda_p m_u}{\sqrt{\pi\varepsilon_0}\,\sqrt{8\pi}\,\pi e^6\alpha\hbar^5 c^2} = \frac{\sqrt{8\varepsilon_0}}{\sqrt{8\pi}\,\pi\alpha^3\hbar^7 c}$$

$$= \frac{\sqrt{8\varepsilon_0}}{\sqrt{8\pi}\,\pi e^2\alpha^2\hbar^6 c^2} = \frac{\sqrt{8\varepsilon_0}}{\sqrt{8\pi}\,\pi e^6\hbar^4 c^4} = \frac{\sqrt{8\varepsilon_0}}{\sqrt{8\pi}\,\pi e^4\alpha\hbar^5 c^3} =^\wedge \frac{\lambda_e R_\infty}{\pi^2 e^2\alpha^4\hbar^6 c^3} = \frac{c\varepsilon_0\,m_e\lambda_e}{\pi^2\alpha^3\hbar^8}$$

$$= \frac{1}{\pi e^6\hbar^3 c^6\,\lambda_p m_u} = \frac{1}{\pi e^2\alpha^2\hbar^5 c^4\,\lambda_p m_u} =^\wedge \frac{1}{\pi e^4\alpha\hbar^4 c^5\,\lambda_p m_u} = \frac{\sqrt{8\pi}\,e^4 c\varepsilon_0}{\sqrt{8\varepsilon_0}\,\pi^2\alpha^5\hbar^9}$$

$$\frac{\varepsilon_0}{\pi^3 e^6\alpha^3\hbar^{11}c^5} \wedge = \wedge \frac{\varepsilon_0}{\pi^3 e^{12}\hbar^8 c^8} = \begin{Bmatrix} 1.\,000,000,013,323,256,090,570,440,076,650,1 \\ 1.\,000,000,013,323,249,553,381,740,475,119,6 \\ 1.\,000,000,013,323,243,016,193,040,873,632,7 \end{Bmatrix}$$

$$= \frac{\varepsilon_0}{\pi^3 e^8\alpha^2\hbar^{10}c^6} \wedge = \wedge \frac{\varepsilon_0}{\pi^3\alpha\hbar^{14}c^2} =^\wedge \frac{\varepsilon_0}{\pi^3 e^4\alpha^4\hbar^{12}c^4} =^\wedge \frac{\alpha\varepsilon_0}{\pi^3 e^{14}\hbar^7 c^9} = \frac{\varepsilon_0}{\pi^3 e^{10}\alpha\hbar^9 c^7}$$

$$= \frac{1}{\sqrt{8\pi}\,\sqrt{8\varepsilon_0}\,\pi^4 e^{10}\alpha\hbar^9 c^{10}} >= \frac{\sqrt{\pi\varepsilon_0}\,\sqrt{8\varepsilon_0}\,\sqrt{8c}\,\sqrt{2c}\,\varepsilon_0}{\sqrt{8\pi}\,\pi^2 e^{12}\hbar^8 c^7}$$

$$\frac{\sqrt{8\pi}\,\pi\alpha^6\hbar^{10}\varepsilon_0^5}{e^{14}c} = \frac{\sqrt{8\pi}\,\pi\alpha^4\hbar^8 c\varepsilon_0^5}{e^{10}} =* \begin{Bmatrix} 1.\,000,001,635,542,503,630,104,648,703,411,2 \\ 1.\,000,001,635,542,497,092,905,344,348,688,9 \end{Bmatrix}$$

$$= \frac{\sqrt{8\varepsilon_0}\,\pi^2\alpha^6\hbar^{10}\varepsilon_0^5}{\sqrt{\pi\varepsilon_0}\,e^{14}c} = \frac{\sqrt{\pi\varepsilon_0}\,\sqrt{8\varepsilon_0}\,\pi\alpha^4\hbar^8 c\varepsilon_0^4}{e^{10}} = \frac{\sqrt{8\varepsilon_0}\,\pi^2\alpha^4\hbar^8 c\varepsilon_0^5}{\sqrt{\pi\varepsilon_0}\,e^{10}} = \frac{\sqrt{\pi\varepsilon_0}\,\sqrt{8\varepsilon_0}\,\pi\alpha^6\hbar^{10}\varepsilon_0^4}{e^{14}c}$$

$$e^{10}\hbar^8 c^8 \wedge =\ \alpha^5\hbar^{13}c^3\,\nu = \qquad = \begin{Bmatrix} 1.\,112,444,508,565,620,169,854,959,028,461,1 \\ 1.\,112,444,508,565,612,897,595,385,589,687,5 \end{Bmatrix}$$

$$e^8\alpha^5\hbar^8 c^9\varepsilon_0^2 \wedge =\ e^4\alpha^7\hbar^{10}c^7\varepsilon_0^2\,\nu = \qquad =* \begin{Bmatrix} 2.\,107,669,048,750,055,537,659,310,981,613,3 \\ 2.\,107,669,048,750,041,759,429,206,563,731,8 \end{Bmatrix}$$

GRAND UNIFIED THEORY MADE EASY

Third edition

Charles R. Storey

Copyright © 2015

published by

Copyright © 2015 by Charles R. Storey

all rights reserved

including the right of reproduction

in whole or in part in any form

ISBN 0-9638766-5-1

Printed in the United States of America